Applications of Calorimetry

Edited by José Luis Rivera Armenta and Cynthia Graciela Flores Hernández

Published in London, United Kingdom

Applications of Calorimetry
http://dx.doi.org/10.5772/intechopen.94666
Edited by José Luis Rivera Armenta and Cynthia Graciela Flores Hernández

Contributors
Rakesh Kumar Soni, Meenu Teotia, Aakansha Sharma, Evgeni Stanev, Maria Dencheva, Luminita Duma, Raquel Gutiérrez-Climente, Elise Prost, Aude Cordin, Carlos Chesta, Viorel Chihaia, Valentin Alexiev, Hasan S. AlMatrouk, Salerwe Mosebi, Pinku Debnath, Krishna Murari Pandey, Balbir Singh Patial, Attila Baranyi, Katalin Kopecskó

First published in London, United Kingdom, 2022 by IntechOpen
IntechOpen is the global imprint of INTECHOPEN LIMITED, registered in England and Wales, registration number: 11086078, 5 Princes Gate Court, London, SW7 2QJ, United Kingdom

British Library Cataloguing-in-Publication Data
A catalogue record for this book is available from the British Library

Additional hard and PDF copies can be obtained from orders@intechopen.com

Applications of Calorimetry
Edited by José Luis Rivera Armenta and Cynthia Graciela Flores Hernández
p. cm.
Print ISBN 978-1-80355-321-4
Online ISBN 978-1-80355-322-1
eBook (PDF) ISBN 978-1-80355-323-8

For EU product safety concerns:
IN TECH d.o.o., Prolaz Marije Krucifikse Kozulić 3, 51000 Rijeka, Croatia,
info@intechopen.com or visit our website at intechopen.com.

Meet the editors

José Luis Rivera-Armenta has a BSc in Chemical Engineering, an MSc in Petroleum Technology and Petrochemicals, and a Ph.D. in Chemical Engineering, all from the Technological Institute of Madero City (ITCM), México. Since 2003, he has been a full-time professor in postgraduate programs at ITCM and head of the thermal analysis, injection, and extrusion laboratory. He has been responsible for several research projects sponsored by Consejo Nacional de Ciencia y Tecnología (CONACYT) and the National Technological Institute of México (TecNM). He has advised bachelor's, master's, and Ph.D. theses. He has published fifty-six scientific articles and six book chapters and has edited three books and one special journal issue. Dr. Rivera-Armenta is also an active reviewer for several journals.

Cynthia Graciela Flores-Hernández received a BSc and MSc in Chemical Engineering from the Technological Institute of Madero City (ITCM), Mexico. She also obtained a Ph.D. in Environmental Sciences from the Autonomous Mexico State University (UAEM). In addition, she completed a postdoctoral stay at the Querétaro Institute of Technology (ITQ), Mexico. She is currently a professor in the Metal-Mechanics Department at ITQ. She has advised bachelor´s and master's students. In addition, she has directed several research projects to respond to national problems. Her research interests include the synthesis and characterization of composite materials using biopolymers, polymer manufacturing by 3D printing, casting, and extrusion.

Contents

Preface

Calorimetry is used to measure the transfer and exchange of heat. It is a technique that has applications in different research and industrial sectors that allows for computing how much heat is generated in a chemical reaction or in the change of state of aggregation.

This book reports on calorimetry in three sections: "Applications in General", "Calorimetry in Materials", and "Calorimetry in Biotechnology".

The first section includes general information on the applications, uses, and characteristics of calorimetry as an evaluation tool. The section presents a general overview of the technique including information on thermodynamic performance analysis in pulse detonation combustors, isothermal calorimetry of molecular interactions between small molecules in organic solvents, and assessment of heat capacity via a thermodynamic approach based on density functional theory calculations.

The second section evaluates the calorimetry technique in different materials. Chapters in this section address such topics as setting time and heat hydration development of Portland cement, using calorimetry to understand structural relaxation in chalcogenide glasses, and cone calorimetry in fire-resistant materials.

The third section focuses on the study of materials from biomolecules related to the skin. This section includes chapters on calorimetry to quantify protein-ligand binding and calorimetry in diagnosing skin allergies.

This book is a useful resource for those interested in understanding the importance and scope of the use of the calorimetry technique.

Dr. José Luis Rivera Armenta
Tecnológico Nacional de México (TecNM),
Instituto Tecnológico de Ciudad Madero,
Ciudad Madero, México

Dr. Cynthia Graciela Flores Hernández
Tecnológico Nacional de México (TecNM),
Instituto Tecnológico de Querétaro
Querétaro, México

Section 1

Applications in General

Chapter 1

Isothermal Calorimetry: Molecular Interactions between Small Molecules in Organic Solvents

Raquel Gutiérrez-Climente, Elise Prost, Aude Cordin, Carlos Chesta and Luminita Duma

Abstract

Isothermal titration calorimetry (ITC) is widely used to study protein-ligand, DNA-drug and/or protein-protein interactions but its application for small molecule complexation remains limited namely when the titration is performed in organic solvents. Compared to other dedicated spectroscopic techniques like nuclear magnetic resonance, infrared spectrometry or fluorimetry, which require a series of experiments to extract site-specific stoichiometry and affinity information, ITC provides in a single experiment a complete thermodynamic picture of the overall interaction mechanism. This chapter presents examples that support the high potential of ITC to probe interactions between small molecules in methanol, acetonitrile and methanol/water mixture on a Nano ITC Low Volume device (TA Instruments), with an emphasis on both simple (1:1) and more complex (1:1 and 1:2) interaction mechanisms.

Keywords: isothermal calorimetry, thermodynamics, association constant, stoichiometry, small molecules, non-aqueous solvents

1. Introduction

Molecular recognition processes, omnipresent in nature, are of crucial importance in all living species. The magnitude of any molecular interaction, which can be translated in terms of heat released or adsorbed, can vary depending on the chemical nature of the interacting partners, on their concentration and the solvation environment in which the process takes place. Developed originally in the middle of the 1960s [1] for studying chemical reactions [2], ITC can measure with high precision and accuracy the heat energy associated with intermolecular reactions but also solvation and dilution experiments. Over the years, ITC became gradually widespread and popular to characterize the thermodynamics signature of molecular interactions in drug design [3, 4], in which the knowledge of the thermodynamic parameters in combination with the structural and kinetic information is decisive during the hit-to-lead optimization, an early step in the drug design process. At this optimization stage, hundreds of compounds with promising affinity against the protein target are screened to identify the best one or two candidate molecules, usually from different chemical series [5]. Compared to other methods employed in the field of drug discovery, ITC offers two clear advantages that facilitates the selection of the candidate molecules: On one side, ITC is the only method that directly measures the reaction enthalpy change [6], which can be considered as an interaction descriptor [7] and therefore an extremely useful parameter for structure thermodynamics correlations. It offers a valuable information notably for the compounds having similar binding affinities but different thermodynamic parameters [7, 8]. Traditionally, drugs characterized by an enthalpy-driven binding were preferred [7, 9]. Nevertheless, the nature of the binding site deserves to be considered because, when an apolar part of the drug interacts with an apolar region of the protein target, the entropic contribution will be favored and the reaction becomes entropy-driven [10]. On the other side, thanks to the detection of the heat change in water molecules and the transfer of protons of the drug molecule during the solvation and dilution experiments, ITC illustrates the differences between polar and apolar interactions that are invisible using techniques such as X-ray crystallography and surface plasmon resonance (SPR) [10]. All these advantages are also highly valuable for other ITC applications such as enzyme-catalyzed reactions [11] and host-guest supramolecular complexation [12]. In supramolecular chemistry for example, ITC combined with supramolecular structure information can provide deeper information on the energies associated with non-covalent interactions (hydrogen bonding, electrostatic, π-π stacking, cation-π and anion-π interactions) and the hydrophobic effect induced by the displacement of water molecules [13].

Recently, the technique appeared also particularly useful and versatile in kinetic assays [14] where the direct measurement of a catalytic reaction [15, 16] was possible. Other applications include the monitoring of microbial activity and dynamics [17, 18], the stability assessment of (bio)pharmaceuticals [19], etc. The development of Low Volume (LV) Nano ITC calorimeters should extend the application fields of this technique not only to biomolecules available in small amounts but also to the study of complexation reactions in organic solvents thanks for example to the availability, on the same calorimeter, of a standard buret handle for aqueous solutions and organic solvents compatible buret handle.

The main advantage of ITC, when compared to other interaction-study dedicated spectroscopies, like nuclear magnetic resonance (NMR), Fourier transform infrared (FTIR) spectrometry or fluorimetry, is the ability to provide in a single experiment the entire thermodynamic profile of the investigated interaction process. In practice, an ITC experiment measures accurately the heat released or adsorbed when a molecule solution is titrated into another in a given aqueous or non-aqueous solution. The large majority of ITC measurements is conducted in aqueous solutions to study protein-ligand, DNA-drug and/or protein-protein interactions [10] whereas most of the ITC investigations in organic solvents or non-aqueous/aqueous mixtures focus on the solvation or dissolution thermodynamic study of various small molecules [20–23], drugs [24], single amino acids [25–27], small peptides [28], metal ions [29], etc. It is worth noting also some complexation studies for copper ions with β-alanine in ethanol [30], 15-crown-5 ether with Na^+ in water-ethanol [31], 18-crown-6 with triglycine in water-acetone and water-dimethyl sulfoxide [32], β-cyclodextrin with benzoic acid in water-ethanol [32], fluorescein isothiocyanate with polymers in water-methanol [33], and the association analysis of urea-based supramolecular polymers in different solvents [34].

The scarce resources in past calorimetry or contemporary ITC literature about intermolecular interactions in organic solvents prompted us to write this chapter.

After a brief introduction of the main basic concepts, we will show the ITC characterization of several 1:1 and 1:2 complexes in organic solvents together with the analysis of the data and the results which provide the thermodynamic profile, stoichiometry and association constant of the reaction. The correlation of ITC results with association or structural information from other spectroscopies may help understanding the formation of even more complex binding events.

2. Isothermal calorimetry: general principles and experimental details

Calorimetry measures the changes in heat released or absorbed during a chemical reaction or a physical process. Heats are measured with instruments called calorimeters classified as adiabatic or isothermal [35]. Heat measured on adiabatic devices show a permanent increase or decrease in temperature. The temperature changes thus obtained are directly related with the heat capacity of the instrument which therefore requires the acquisition of distinct experiments for calibration. In isothermal calorimeters the heat is allowed to flow between the reaction cell and a heat sink surrounding the two cells (see **Figure 1a**) and is actively regulated to maintain a constant level by power compensation. The development of micro- and nano-calorimetry rendered possible the detection of very small heat changes in small volumes of samples. Nano calorimeters have a detection limit in the nanowatt range [35]. Historically, Nano calorimeters have been applied to the study of the reactants available in reduced amounts like biomolecules.

The design of the NanoITC from TA Instruments relies on the differential power compensation technology (see **Figure 1a**) which allows to optimize sensitivity and responsiveness. Nanowatt sensitivity is achieved thanks to an internal reference. Two cylinder-shaped identical chambers (also called sample and reference cells) of 170 μL are located in a compartment which works as a thermal barrier. As described in the user manual [36], semiconducting thermoelectric devices (or TED) control and detect temperature differences between the sample and the reference chambers. In a titration experiment, both cells are entirely filled: the reference cell with the pure solvent, the sample cell with one reactant (or titrand) and the syringe with the other reactant (or titrant) of the reaction under study, both in the same solvent. The titrant is usually prepared at 10-fold higher concentration for a 1:1 binding model [37]. ITC titration experiments are implemented by incremental injection of a precise volume of titrant into the solution of titrand at discrete time intervals (see **Figure 1c**). Typically, about 25 injections per experiment are performed using a motor-driven syringe capable to deliver defined volume within 1–10 μL per injection. The syringe is coaxially introduced in the sample cell through a long access tube. The stepping motor precisely controls not only the injection volume but also the stirring speed of the reactants in the sample cell. Therefore, for each injection, the interaction between the two reactants releases (or adsorbs) heat that increases (or decreases) the sample cell temperature. This temperature change will activate the feedback heat controller (power compensation) on the sample cell such as to maintain a zero temperature difference between the two cells. For each heat variation during a stepwise titration, the feedback regulator will compensate this difference by decreasing (or increasing) the heat of the sample cell by the amount of heat supplied by the reaction and, at each variation associated to each injection step, will lead to a peak in the thermogram. **Figure 1c** illustrates the construction of a representative thermogram as the stepwise titration proceeds. During the titration experiment, the reactant in the sample cell is gradually

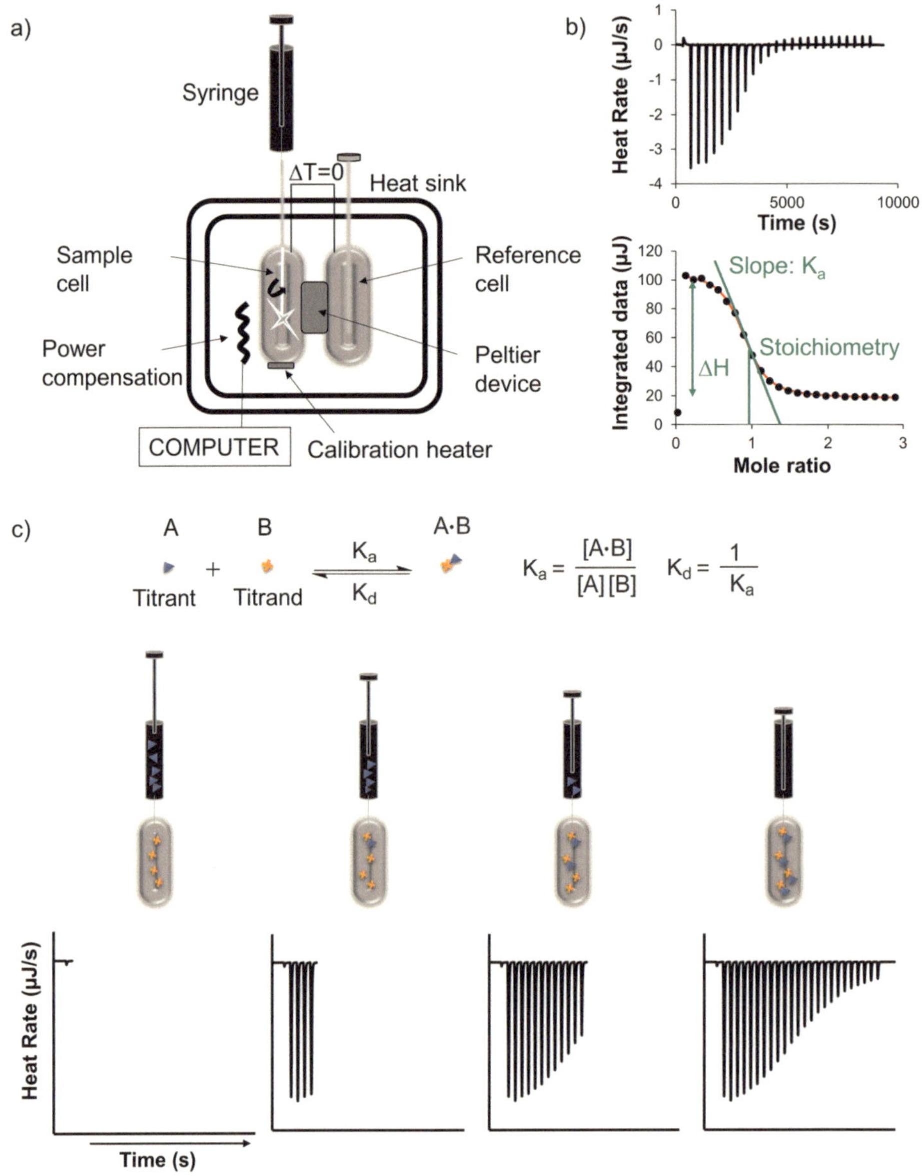

Figure 1.
a) Drawing of the Nano ITC measuring unit (TA instruments) and its basic elements. b) Typical stepwise ITC raw thermogram and the corresponding integrated data. The parameters obtained after adjusting the data with the one-site binding model are also highlighted in green. c) Chemical reaction and illustration of the evolution of the stepwise raw thermograph as the titrant is injected in the sample cell.

transformed into the molecular complex. The injection of the titrant is conducted until the titrand in the sample cell is fully saturated and the heat signal becomes equal to the background heat generally equal to the dilution heat of the titrant [37].

The normalized integrated area of each peak is next approached with the appropriate model to estimate the affinity, enthalpy and stoichiometry of the interaction. The first recorded experiment can be further analyzed with the Experiment Design tool of NanoAnalyze to find out the optimal concentration conditions leading to a "S-shape" thermogram (see **Figure 1b**). The titration experiment is then repeated with these optimized concentrations and the data analyzed. The integration of each peak area in the thermogram (**Figure 1b** top) gives the amount of heat exchanged while injecting a known amount of the reaction partner into the sample cell. When the concentrations in the syringe and the sample cell were appropriately chosen, a display of the integrated heat signal as a function of mole ratio (of the injected compound over the one contained in the sample cell) reveals a typical sigmoidal shape (**Figure 1b** bottom). An inspection of the integrated titration curve gives information on the molar enthalpy (ΔH) from the height of the curve, the number of binding sites (or stoichiometry) of the reaction from the position of the inflection point on the mole ratio axis, and the association constant (K_a) from the slope of the curve. Nowadays, the companies provide an analysis software that incorporates the specifics of the instrument. It is therefore relatively easy to make a first estimation of the ΔH, n and K_a quantities thanks to various models that approach the experimental data using a nonlinear routine to find the most probable thermodynamic parameters describing the interaction process.

In the absence of any solubility or aggregation issues, direct and reverse titration [38, 39] can be considered when studying the interaction of small molecules in a given solvent. For two A and B interacting molecules (**Figure 1c**), the injection of aliquots of A into B is called direct titration whereas the injection of aliquots of B into A is termed reverse titration. As previously mentioned, the titration experiment is well performed when the concentrations are chosen such as the molecule in the sample cell is fully saturated. Assuming that the molecular mass of reactants does not affect the equilibrium binding equations (microscopy reversibility), direct and reverse titrations should be equivalent. They can be modeled by the same set of equations and should lead to the same thermodynamic parameters for a reversible reaction.

The reaction in **Figure 1c** corresponds to a complex where a single molecule A interacts with a single molecule B (i.e., the one to one model which corresponds to a stoichiometry of 1). Since the reaction enthalpy change (ΔH) and association constant (K_a) are directly measured from an ITC titration experiment, under constant temperature and pressure, the Gibbs free energy (ΔG) and the entropy (ΔS) can obtained using the following relationships:

$$\begin{cases} \triangle G = \triangle H - T\triangle S \\ \triangle G = -RT\ln K_a \end{cases} \tag{1}$$

with R the ideal gas constant (R = 8.314 J/mol·K) and T the temperature in K. Equilibrium association constants can be obtained with acceptable statistical precision if the following condition is fulfilled:

$$1 < c < 1000 \tag{2}$$

where c (also known as c-value) is given by the relationship $c = nK_a[B]$, [B]: concentration of B molecule [40]. The condition in Eq. (2) is directly related to the shape of the binding curve (also termed thermogram). For protein-ligand complexes generally characterized by an affinity in the μM range (i.e. 10^6 M^{-1} Ka), a c-value within 20 and 200 is recommended in order to minimize K_a errors and thermodynamic parameters as shown by comparing ITC reports from different laboratories on the same benchmark protein-ligand complex in buffer [41–43].

By performing the ITC titration experiment at different temperatures, the change in heat capacity, ΔC_p, for the A·B complex formation can be estimated according to the following equation:

$$\Delta C_p = \frac{d\Delta H}{dT} \tag{3}$$

with ΔC_p in J/mol·K.

The enthalpy represents the energy change of the system when the molecule A interacts with B in a given solvent. Different type of noncovalent interactions (hydrogen bonds, ions pairs, van der Waals forces, etc.) can take place at the binding interface and therefore affect the enthalpy change. For example, the formation of noncovalent interactions between atoms is an exothermic process characterized by a negative enthalpy change whereas their breaking is an endothermic one with a positive enthalpy change. The heat released during a binding process describes the entire system under study with individual contributions from the interacting partners but also the solvent. In reality, the measured enthalpy change upon binding is the sum of many positive and negative ΔH contributions resulting from the simultaneous formation and disruption of noncovalent interactions [44]. Like in protein-ligand interactions in an aqueous medium, the observed ΔH of binding in an organic solvent is a global property which reflects the partial loss of solvent contacts of the interacting partners, the formation of complex noncovalent interactions and the solvent rearrangement near the complex surface.

As main direct experimental observable in calorimetry, the measured heat is correlated with the reaction taking place at the molecular level and the aim of the calorimetry is to provide reliable heat data capable to characterize molecular interactions. In the literature, the enthalpies directly obtained by calorimetry have been correlated with the values of binding enthalpy derived from the van't Hoff relationships. Experimental and simulation studies have shown that statistically relevant discrepancies are notably found when the experimental setup or data analysis are not correctly performed [45, 46].

Incremental titration described previously is the most common titration method used. Continuous titrations, which consists of constantly injecting the titrant into the calorimeter vessel while monitoring the thermal power, are shorter than the incremental ones and therefore, they can be of great interest for unstable samples. The development of the continuous ITC (cITC) method for micro- and nano-calorimeters [47] rendered the technique even faster and more versatile for the study of thermodynamic processes in a complex interaction. Interestingly, it can also represent a quick alterative to find out the concentration conditions leading to an exploitable thermogram. The screening of the optimum conditions (Eq. 2) may be speeded up even more if the cITC thermograms can be exploited by the Experiment Design tool in NanoAnalyze software, as mentioned for the classical ITC data. Unfortunately, this is not yet the case with the currently available NanoAnalyze software (3.12.0).

Another advantage of the cITC, is the potential expansion of the equilibrium constants accessible by ITC. Indeed, Markova and Hallén [47] have shown by computer simulations that cITC expands by 3 orders of magnitude the range of K_a achievable by ITC. Therefore, for cITC the Eq. (2) becomes:

$$1 < c < 3x10^6 \tag{4}$$

with 10^{12} M^{-1} being the highest equilibrium constant reachable by cITC. This is rendered possible by the increased data points density in a continuous titration experiment which better defines the slope region in a 1:1 binding curve.

3. Experimental details

The reagents used in this work are summarized in **Table 1**. All measurements were performed on a differential power compensation Nano ITC Low Volume (Nano ITC LV) calorimeter from TA Instruments (Waters, France), using a 50 μL injection syringe while stirring at 400 rpm (the maximum stirring value). The Nano ITC LV has two gold 170 μL reaction vessels and a buret assembly holding a stainless-needle syringe with a twisted paddle at the tip and titrant exit at the bottom. Therefore, the syringe serves not only for the delivery of the titrant but also for stirring. To avoid the formation of bubbles in the cells and syringe, the samples were degassed in a vacuum degassing station [48] for 15 minutes immediately before use. Injections were started after achievement of baseline stability (using the automatic equilibration mode for "small Heats" of ITCrun program controlling the calorimeter with the following criteria: absolute acceptable slope: ΔH 0.1 μW/h; acceptable absolute standard deviation: 0.01 μW). In addition, an equilibration time of 300 s has been considered before the first and after the last injection to assess the quality of the baseline. The experimental parameters were: 350 μL in the sample cell, 50 μL in the syringe, 350 μL solvent in the reference cell (changed weekly), 25 injections of 2.02 μL except for the first injection which was of 0.48 μL. The integrated heat effects of each injection were corrected by subtraction of the corresponding integrated heat effects associated with titrant dilution into the solvent. The experimental data obtained from the corrected calorimetric titration were analyzed on the basis of different interaction models with the NanoAnalyze software. The first injection was not taken into consideration for data analysis.

The electrical calibration of the calorimeter was performed according to the manufacturer's instructions. Water in water dilution experiments are regularly performed to check and validate the initial criteria of the manufacturer. Cells and syringe correct cleaning is essential to avoid artifacts and produce good quality data. Additionally, it is crucial to keep the needle of the syringe perfectly straight. Cleaning of the sample cell can be performed automatically using the vacuum of the degassing device and is generally done with 1 L of 2.5% DECON followed by 1 L of MilliQ water. An ITC

Name	Purity %	Supplier	Reference	CAS number
Acetonitrile	99.95	Biosolve	UN1648	75–05-8
Benzoic acid	≥ 99.5	Sigma-Aldrich	242,381-25G	65–85-0
DMAPMAm[a]	99	Aldrich	409,472-250ML	5205-93-6
Glucuronic acid	≥ 98	Sigma	G5269	6556–12-3
Isophthalic acid	99	Alfa Aeser	A14445	121–91-5
Methanol	100	VWR	20,847–320	67–56-1

[a]*N-[3-(dimethylamino)propyl]methacrylamide.*

Table 1.
Reagents used in this work.

experiment takes between 1 and 2 hours depending on the injection delay and the equilibration duration necessary to fulfill the heat stability statistical criteria. ITCRun, the software which controls the Nano ITC calorimeter, does not record the evolution of the heat values during the equilibration delay but only its duration.

Despite the relative simplicity of ITC experiments, the selection of the right binding model for the fitting of the experimental data in order to estimate the thermodynamic parameters can be challenging, especially for complexes where one of the reactants present multiple sites and there is not previous information about the stoichiometry or binding mechanism [49]. Among the various softwares currently available for the titration data analysis, the softwares provided by the ITC manufacturers, i.e. Origin from MicroCal/Malvern and NanoAnalyze in the case of TA instruments, the interpretation of the binding isotherms can be done for each data set individually using either classical models such as one-site independent or two-sites sequential models, multiple sites, dimer dissociation, cooperative and competitive replacement models. The possibility of dilution subtraction or the selection of a control model together with the binding model remains at the user's choice.

Over the last decade, alternative softwares have been developed for more complex processes, e.g. AFFINImeter [50], pytc [51], HypΔH [52], CHASM [53] or SEDPHAT [54]. The free platform SEDPHAT gives the possibility to combine several experimental data (calorimetry, spectrophotometry, sedimentation and surface binding assays) in a single global analysis. The main purpose of the platform development was to reduce the discrepancies between the thermodynamic parameters obtained using different devices and setups (e.g. the use of different sample volume, concentrations, immobilization of the template, etc...). The SEDPHAT results presented in the present chapter concern only calorimetry data in order to perform a global analysis of several measurements, including repetitions of the performed direct and reverse titrations, increasing therefore the confidence in the binding parameters obtained individually with NanoAnalyze software.

4. ITC data and their evaluation

The following subsections illustrate the extraction of thermodynamic parameters and association constant from the experimental raw data for one-site and two-sites binding complexes in different non-aqueous solvents.

4.1 1:1 Complexes in different solvents and calorimetric heat capacity

Molecular recognition processes are the archetypal reactions in various domains going from life science to technology and the design of a suitable target which binds the other partner with specificity remains challenging nowadays. The study of the binding complexes presented herein was originally motivated by the need to optimize the design of molecularly imprinted polymers [55] by characterizing the affinity and the complete thermodynamic profile of the monomer-target interaction in the solvent used for the synthesis of the final polymer. A better comprehension of the interactions and energies involved in the preorganization of the monomers around the target could improve their synthesis protocols by selecting monomers with high affinity for the target.

The feasibility of elucidating the thermodynamic parameters for a 1:1 binding model using stepwise titration experiments is first demonstrated with the interaction

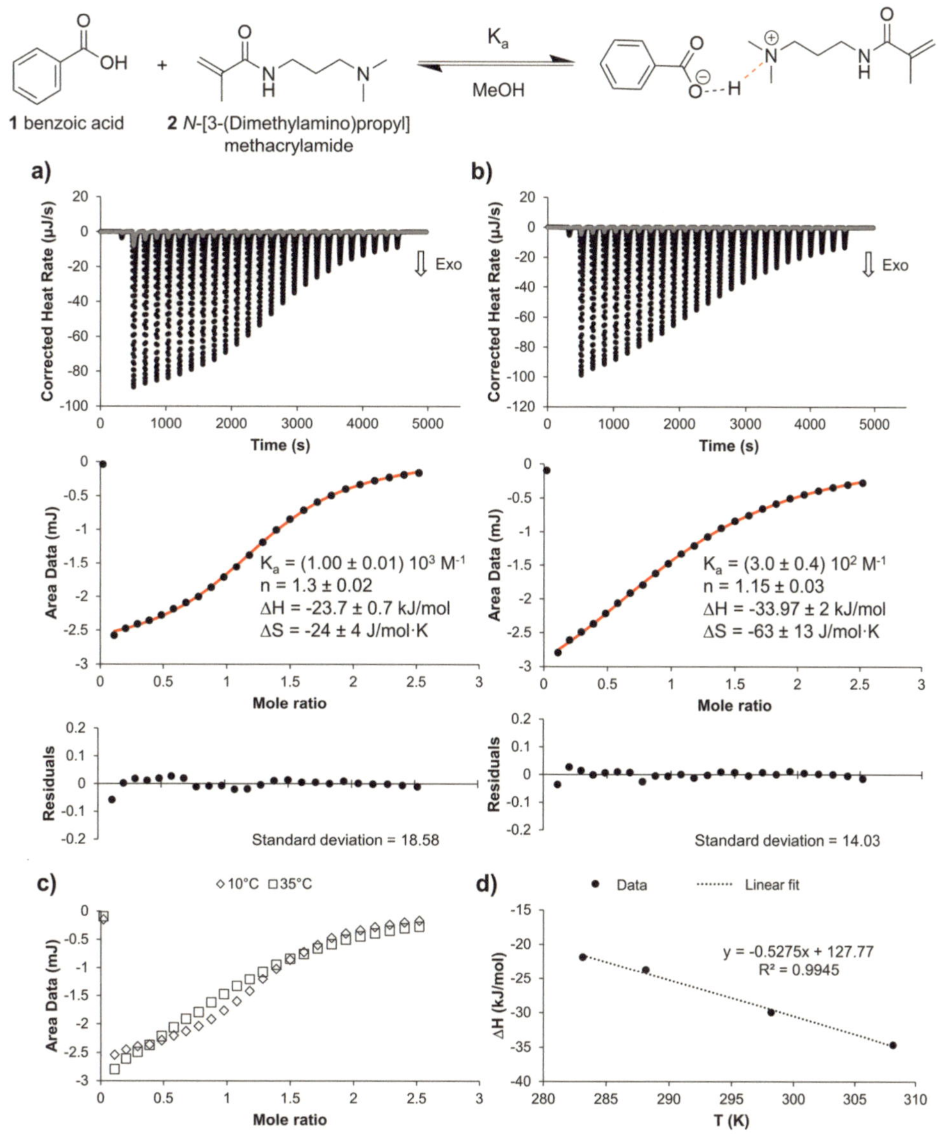

Figure 2.
ITC titration of **2** *into* **1** *in MeOH at 15 (a) and 35°C (b) as filled black circles. The dilution of* **2** *into MeOH is also shown as gray filled circles. The middle thermogram shows the integrated heat data as filled black circles and their nonlinear fitting using the 1:1 binding model as red continuous line. c) Integrated corrected data at 10 and 35°C. d) Linear fit of the calorimetric ΔH values versus temperature which gives access to the change in heat capacity,* $\Delta C_p \sim -0.5$ *kJ/mol·K.*

between benzoic acid (**1**), a food antimicrobial, and *N*-[3-(dimethylamino)propyl] methacrylamide (**2**), a monomer, in methanol (**Figure 2**). In water at 25°C, **1** shows a pKa of ~4.2 [49].

whereas **2** is a typical amine base characterized by a pKa of ~9.2. In methanol and other organic solvents, the pKa cannot be defined but it is expected that **1** and **2** behave as a strong Lewis acid and base, respectively. Qualitative evidence of the interaction between benzoic acid and 2-dimethylaminoethyl methacrylate-based copolymers has been provided by infrared (IR) and ^{1}H NMR spectroscopies [50].

^{1}H NMR spectra recorded for different molar compositions of benzoic acid:copolymer containing amine groups in acetone showed that carboxyl-amine interaction causes a shift of the aromatic proton resonances to high field (small ppm values) whereas protons near the amine group shift to low field (big ppm values). These chemical shifts suggest that the carboxyl-amine complexation is associated with an increase in electron density for the carboxyl and a decrease of the electron density for the amine group, results consistent with the formation of a contact ion pair stabilized by attractive electrostatic forces and H-bonding. The same conclusions are reached by analyzing the IR spectra of the mixtures in acetone. For these complexes in chloroform and acetone solvents, it was stated that the excess of benzoic acid tends to self-association producing therefore benzoic acid dimers in addition to the benzoic acid-amine complex.

Figures 2 and **3** display the calorimetry data for the interaction between benzoic acid (**1**) and *N*-[3-(dimethylamino)propyl]methacrylamide (**2**) in methanol and acetonitrile, respectively. In order to estimate the heat capacity changes, the heat reaction response has been measured in methanol over a temperature range between 10 and 35°C. The top panels in **Figure 2a** and **b** indicate that the complexation is exothermic over the full temperature range spanned. The middle and the bottom panels show the integrated experimental heat data fitted with the 1:1 independent binding model and the residuals which describe the differences between the interaction model and the measured data. In all experiments, 60 mM **2** were titrated into 8 mM **1** except

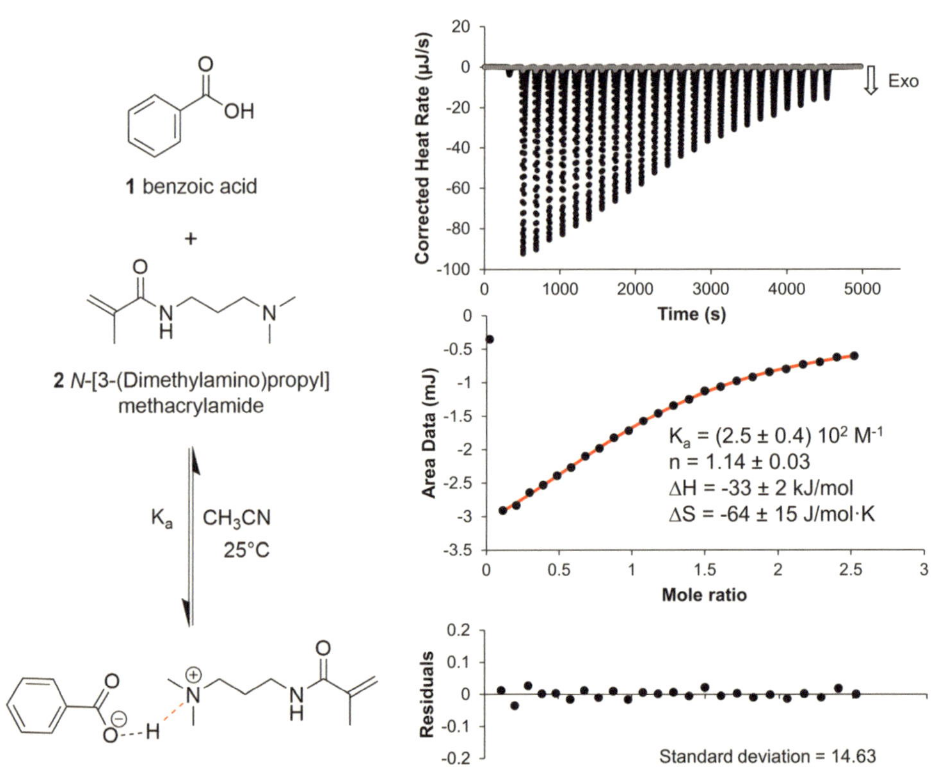

Figure 3.
*ITC titration of **2** into **1** in acetonitrile at 25°C as filled black circles. The dilution of **2** into acetonitrile is overlaid as gray filled circles. The middle thermogram shows the integrated heat data as filled black circles and the nonlinear fitting of the data using the 1:1 binding model as red continuous line.*

T °C	K_a M^{-1}	n	ΔH^a	ΔS^a	ΔG^a	c^b
10	$(1.5 \pm 0.1)\ 10^3$	1.36 ± 0.02	-21.3 ± 0.4	-14 ± 4	-17.3 ± 0.7	16
	$(1.3 \pm 0.1)\ 10^3$	1.30 ± 0.02	-22.4 ± 0.5	-19 ± 5	-17.0 ± 0.8	14
15	$(1.00 \pm 0.01)\ 10^3$	1.30 ± 0.02	-23.7 ± 0.7	-24 ± 4	-16.7 ± 0.1	10
25	$(6.7 \pm 0.9)\ 10^2$	1.07 ± 0.02	-30 ± 1	-46 ± 9	-16 ± 1	6
	$(5.8 \pm 0.3)\ 10^{2c}$	1.07 ± 0.01^c	-21.9 ± 0.4^c	-20 ± 4^c	-15.9 ± 0.5^c	6^c
	$(7.0 \pm 0.1)\ 10^{2d}$	1.16 ± 0.03^d	-25 ± 1^d	-28 ± 7^d	-17 ± 1^d	6^d
	$(7.0 \pm 0.1)\ 10^{2d}$	1.13 ± 0.04^d	-26 ± 2^d	-32 ± 9^d	-17 ± 1^d	6^d
	$(6.4 \pm 0.6)\ 10^{2e}$	1.1 ± 0.6^e	-30.0 ± 0.5^e	-46 ± 8^e	-16.1 ± 0.9^e	-
	$(2.5 \pm 0.4)\ 10^{2f}$	1.14 ± 0.03^f	-33 ± 2^f	-64 ± 15^f	-14 ± 2^f	2^f
35	$(3.0 \pm 0.4)\ 10^2$	1.15 ± 0.03	-34 ± 2	-63 ± 13	-15 ± 2	3
	$(2.7 \pm 0.3)\ 10^2$	1.15 ± 0.03	-35 ± 2	-67 ± 14	-14 ± 1	2

[a]*kJ/mol, ΔS is given in J/mol K.*
[b]*calculated c-value.*
[c]*100 mM 2 into 10 mM **1**.*
[d]*60 mM **1** into 8 mM 2 (reverse titration).*
[e]*global analysis done on direct and reverse datasets and performed with SEDPHAT platform [54], n value is calculated based on the values obtained in the individual analysis with NITPIC [56] as SEDPHAT employs a different parameter in its analysis.*
[f]*in acetonitrile.*

Table 2.
Best-fit thermodynamic parameters from ITC measurements for the 1–2 complex in methanol.

otherwise indicated. **Table 2** summarizes the ensemble of thermodynamic parameters together with the stoichiometry and the association constant derived using a 1:1 independent binding model at each temperature. The calculated *c*-value is also shown and falls in the range 2–20. The concentration conditions have been optimized using the tool "ITC Experiment Design" from NanoAnalyze at 25°C for a *c*-value ~10. A greater *c*-value would have required higher concentrations for benzoic acid and therefore would have prevented us from carrying out the reverse titration owing to benzoic acid reduced solubility in MeOH above 60 mM. In parallel, to check the accuracy of the **1–2** affinity and thermodynamics parameters obtained by ITC, a global fit analysis for both direct (60 mM **2** into 8 mM **1**) and reverse (60 mM **1** into 8 mM **2**) titrations in MeOH at 25°C has also been performed with the open source SEDPHAT platform. First, the ITC data have been integrated using the NITPIC software which also allows to subtract the corresponding dilution for each direct and reverse titration dataset and gives access to the stoichiometry parameter. Second, the direct and indirect NITPIC datasets were integrated and saved in a SEDPHAT configuration file. The data were then fitted with the one-site binding model and the statistics of the thermodynamic parameters calculated for a confidence level at 95%. The results obtained with SEDPHAT (**Table 2**) are almost identical to the ones obtained for the direct titration with NanoAnalyze. SEDPHAT contains explicit factors which can account for errors in the active concentrations. Their inspection for the individual analysis of both and direct titration data suggests an error in concentrations notably for the indirect titration. This might explain the bigger estimations of the stoichiometry and enthalpy change in the case of reverse titration.

The results summarized in **Table 2** show that the **1–2** complexation is an exothermic ($\Delta H < 0$) and enthalpy-driven ($|\Delta H| > |T\Delta S|$) process at all temperatures and experimental conditions studied. K_a decreases with increasing temperature while ΔH becomes more exothermic as the temperature rises. The $|\Delta S|$ show also a tendency to increase with temperature. Within the experimental uncertainties, the calculated |ΔS| values, which range within 20 and 50 kJ/mol·K, are in good agreement with the expected entropy change characterizing the ion pair formation from neutral reactants in polar solvents [57, 58]. The decrease of the association constant by about one order of magnitude between 10 and 35°C is particularly relevant for the design of molecularly imprinted polymers because their synthesis is usually performed at temperatures higher than the ones used here (50 to 70°C). It therefore suggests the importance of probing the interaction between the target and the functional monomer at the synthesis temperature.

In order to estimate the heat capacity changes, the heat reaction response has been measured over a temperature range between 10 and 35°C. The analysis of the variation of ΔH as function of T (**Figure 2d**) allows to estimate a value for $\Delta C_p \sim -0.5$ kJ/mol·K. Formally, for a simple equilibrium studied over a relatively narrow temperature range, ΔCp should be zero. Thus, these results suggest that the complexation process may be more complicated than initially assumed. Taken together, these results suggest that the **1–2** complex formation mechanism probably involves two or more coupled equilibria whose relative importance depends on temperature and reactant concentrations. For example, if at the concentrations used, a small fraction of the benzoic acid is in the dimeric form, the **1–2** acid/base complex formation requires the dissociation of the dimer and therefore a coupling of the two equilibria (complexation and dimerization). This can explain the discrepancies observed in n (which is clearly different of 1), ΔH and ΔS calculated at different temperatures (**Figure 2a** and **b**). Coupled equilibria may also explain the differences between "direct" and "reverse" titrations.

Figure 3 shows ITC titration of 60 mM of **2** in acetonitrile in 8 mM of **1** in the same solvent at 25°C. The thermodynamic parameters obtained from fitting the experimental data are listed in **Table 2**. This study was performed to investigate the role of H-bond interactions in the thermodynamics of the reaction. Acetonitrile and methanol show similar dielectric constants (36.6 and 33.0, respectively) and thus should exhibit similar abilities to stabilize the ionic product (relative to the reactants). However, both solvents differ markedly in their ability to form H-bonds. Methanol, a protic solvent, should further stabilize reactants and reaction product through the formation of such bonds. However, a comparative analysis of the ΔG values (ΔH and ΔS) obtained in acetonitrile and methanol at 25°C (see **Table 2**) shows that, within experimental uncertainties, they are practically identical. These results suggest that H-bonds do not preferentially stabilize neither the reactants nor the complex, and therefore, they are not determinant in the thermodynamics of the complexation process.

The second complexation process explored by stepwise ITC titration concerns the interaction between the glucuronic acid (**4**), a sugar acid derived from glucose, and (4-acrylamidophenyl)amino methaniminium acetate (**5**), a polymerizable benzamidine salt, which can form with carboxylates stoichiometric non-covalent complexes characterized by affinities higher than 10^3 M^{-1} [59]. We previously studied the **4–5** complexation by ^{1}H NMR spectroscopy in DMSO-d_6 [60]. Job's plot [61, 62] and titration experiments demonstrated the formation of a 1:1 complex with an affinity of 7.1 10^3 M^{-1}. The complex formation in MeOD/D_2O (4/1 v/v) gives slightly smaller chemical shift differences and an association constant of 4.4 10^3 M^{-1}.

Figure 4 displays the calorimetry data obtained at 25°C in MeOH/H_2O (4/1 v/v) for direct (i.e., monomer into the glucuronic acid) and reverse (i.e., glucuronic acid

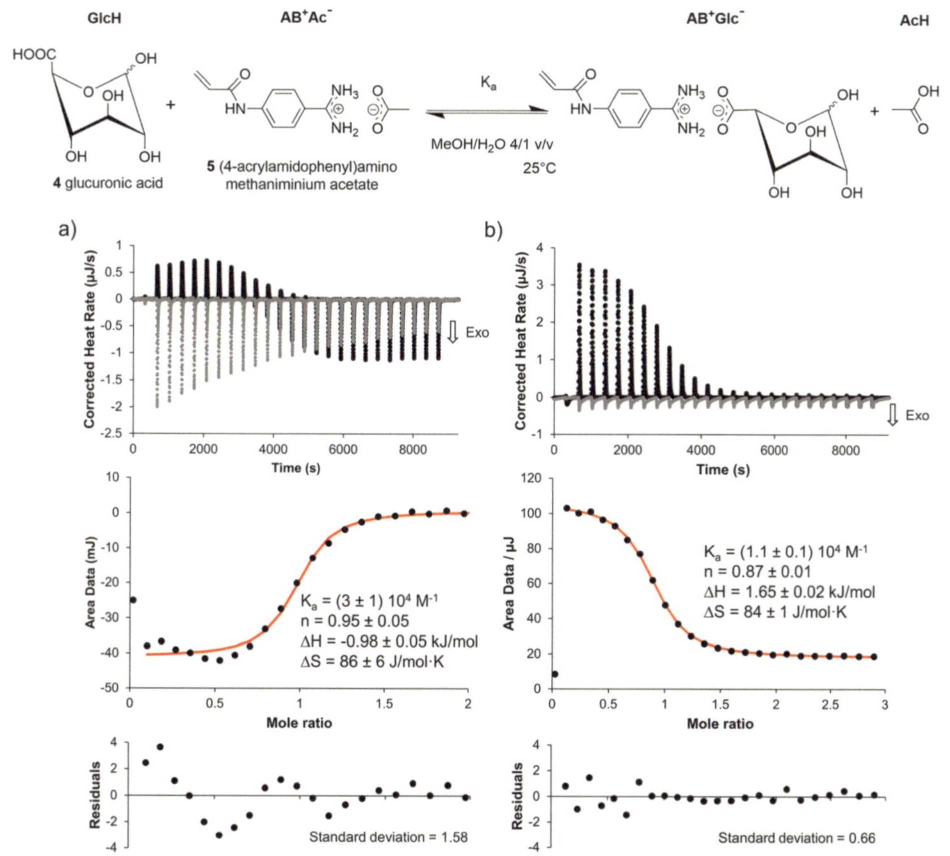

Figure 4.
*Direct (a) and reverse (b) titrations of AB^+Ac^- (5) into GlucH (4) in $MeOH/H_2O$ (4/1 v/v) at 25°C as filled black circles. The dilutions of **4** and **5**, respectively, into $MeOH/H_2O$ (4/1 v/v) are also shown as gray filled circles. The middle thermogram shows the integrated heat data as filled black circles and the nonlinear fitting of the data using the 1:1 independent binding model as red continuous line.*

into the monomer) titrations. An inspection of the reaction scheme (top of **Figure 4**) suggests that the process can be considered as a displacement reaction (i.e., a reaction where glucuronic acid replaces acetate as a counter ion of the salt). The titration thermograms in top panels of **Figure 4a** and **b** show that the macroscopic experimental heat either changes from endothermic to exothermic (direct titration) or remains

K_a M^{-1}	n	ΔH[a]	ΔS[a]	ΔG[a]	c[b]
(3 ± 1) 10^4[c]	0.95 ± 0.05[c]	−0.98 ± 0.05[c]	86 ± 6[c]	−26 ± 2[c]	86 [c]
(1.1 ± 0.1) 10^4[d]	0.87 ± 0.01[d]	1.65 ± 0.02[d]	84 ± 1[d]	−23.3 ± 0.4[d]	29[d]

[a]*kJ/mol, ΔS is given in J/mol K.*
[b]*calculated c-value.*
[c]*21 mM **5** into 3 mM **4** (direct titration).*
[d]*25 mM **4** into 3 mM **5** (reverse titration).*

Table 3.
*Thermodynamic parameters of the **4–5** complex in $MeOH/H_2O$ (4/1 v/v) at 25°C.*

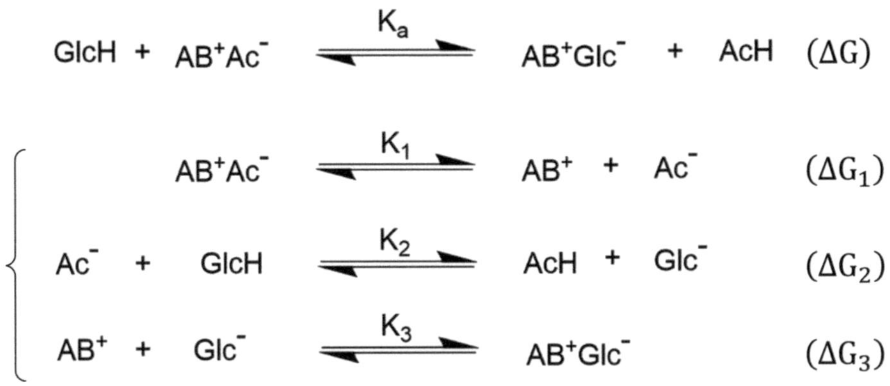

Figure 5.
Global displacement reaction described as a 3 steps chemical process.

endothermic (reverse titration). The dilution of **4** and **5** is exothermic. The nonlinear fitting of the macroscopic heat data after dilution subtraction gives the thermodynamic parameters, the stoichiometry and the association constant reported in **Table 3**. The relatively small errors can be correlated with the calculated *c*-value, which is an order of magnitude higher than for the **1–2** complex described previously.

For both (direct and reverse) titrations, the enthalpy change (ΔH) is close to zero whereas the Gibbs free energy change (ΔG) is negative, which indicates a spontaneous process. Since $T|\Delta S| >> |\Delta H|$, the reaction is entropy driven. As we show below, the negative driving force of the displacement reaction is mainly due to the observed differences in the association constant (K') of the two acids. In water, the reported pKa are 4.75 [63] and 2.88 [64] for acetic and glucuronic acids, respectively. This difference in acidity is explained by the presence of one oxygen atom in the carbon α to the glucuronic -COOH group, which significantly weakens the O-H bond. The thermodynamics of the displacement reaction can be analyzed as a reaction occurring in consecutive stages. This is possible because Gibbs free energy is a state function of the system (i.e., it is independent pathway taken but only on the initial and final states) and therefore, the global displacement reaction (top of **Figure 5**) can be written as the sum of 3 hypothetical steps reaction characterized by their Gibbs free energies: ΔG_1, ΔG_2 and ΔG_3, so that $\Delta G = \Delta G_1 + \Delta G_2 + \Delta G_3$.

The first step describes to the dissociation of the AB^+Ac^- into the corresponding solvated free ions characterized by $\Delta G_1 > 0$. The second step is an acid-base reaction in which the acetate and the glucuronic acid exchange a proton for which the ΔG can be calculated as explained latter. Finally, the third step corresponds to the association of the glucoronate ion with the monomer counter ion to complete the displacement process and has a $\Delta G_3 < 0$. Although ΔG_1 and ΔG_3 are not identical, as they characterize the dissociation/association of different ion pairs, they will have similar orders of magnitude and we can assume that: $\Delta G_1 + \Delta G_3 \sim 0$. Based on this hypothesis, the main contribution to the global reaction should come from ΔG_2. The equilibrium constant for this process (K_2) can been determined from the pKa values to be $\sim$80. Using the relationships of Eq. (1), we obtain $\Delta G_2 \sim -10$ kJ/mol. Although this value is half of that obtained experimentally, it should be remembered that the K_2 was roughly calculated from the pKa obtained in water. The decreased dielectric constant in the MeOH/H_2O (4/1 v/v) mixture used in this study should enhance the differences in acidity between the two acids, making the process even more spontaneous.

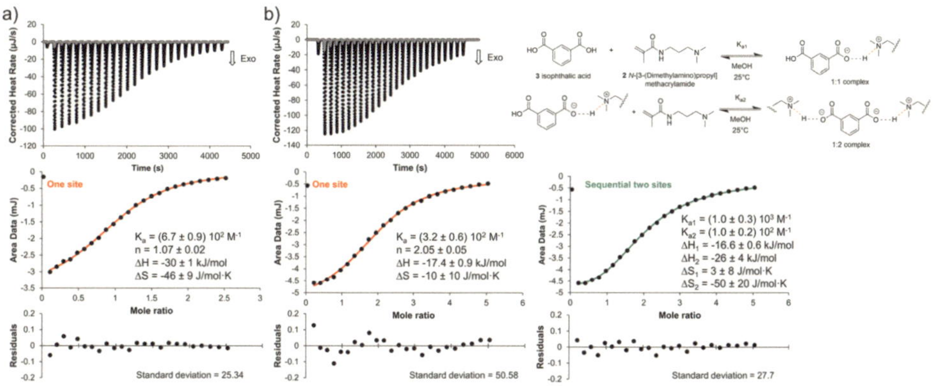

Figure 6.
ITC titration for 1:1 and 1:2 complexes in methanol at 25°C. a) 2 into 1 and b) 2 into (3) titrations as black filled circles. The dilution of 2 into methanol is shown as grey filled circles on the top graphs as well. The middle thermogram shows the integrated heat data as filled black circles and the nonlinear fitting of the data using the one-site binding model as red continuous line or the sequential two-sites binding model for the interaction with isophthalic acid (right side) as green continuous line.

4.2 Two-sites complexes

The involvement of more than one molecular process into a reaction is not always directly visible on the measured heat profile. The casual case where the two interaction mechanisms have an enthalpy that nicely emphasizes each inflection of 1:2 (or 1:3) complexation is relatively rare. **Figure 6** shows the example of a two-sites assembly with the interaction between the dimethylamino-based monomer and isophthalic acid, a target with two carboxylic groups, by putting it in mirror of the complex formed between the same monomer with benzoic acid. Both titrations have been carried out at 25°C in methanol.

The experimental heat response for a 2:1 complexation could be in theory approached with three different binding models (one-site independent, two-sites independent and two-sites sequential). Nevertheless, different arguments may help to restrict the plausible models: close inspection of the data to see if the two individual binding processes are clearly visible on the heat response, choice of the model with the smallest number of adjustable parameters, analysis of the nature of the functional groups of the interacting partners, interaction knowledge from other techniques, etc.

For all the thermograms in **Figure 6**, the binding isotherms were obtained as previously described by integrating the individual heat variation which results from the injection of the titrant. In both cases, initially a nonlinear least-squares regression analysis considering a single binding site was employed for both titrations. As represented in **Figure 6a**, the model fits well the data for the complex with benzoic acid, allowing to estimate the dissociation constant and the binding enthalpy while the stoichiometry value close to 1 confirmed the one-site interaction. In the case of the complex with the isophthalic acid, the one-site binding model is the simplest one to test with only 3 variable parameters and could represent a first acceptable choice if we assume the two carboxyl groups as chemically equivalent. However, this assumption is not strictly true, since it has been reported that the pKa values of the two carboxyl groups of **3** at 25°C in water are: pK_{a1} = 3.46, pK_{a2} = 4.46 (ref). Therefore, this model is expected to estimate a stoichiometry of 2 together with a binding enthalpy (ΔH) and an association constant (K_a) as an average of the two complexation events. **Table 4** collects the thermodynamic parameters

	$K_a\ M^{-1}$	n	ΔH^a	ΔS^a	ΔG^a	c^b
1:1	$(6.5 \pm 0.9)\ 10^2$	1.07 ± 0.02	-30 ± 1	-46 ± 4	-16 ± 1	6
2:1[c]	$(3.2 \pm 0.6)\ 10^2$	2.05 ± 0.02	-17.4 ± 0.9	10 ± 10	-14 ± 2	7
2:1[d]	$(1.0 \pm 0.3)\ 10^3$ $(1.0 \pm 0.2)\ 10^2$	1 1	-16.6 ± 0.6 -26 ± 4	3 ± 8 -50 ± 20	-17 ± 2 -11 ± 4	—

[a]*kJ/mol except for ΔS which is given in J/mol K.*
[b]*calculated c-value.*
[c]*one-site independent binding model.*
[d]*sequential two-sites binding model.*

Table 4.
Thermodynamic parameters obtained by nonlinear fitting of heat data corresponding to the titration of 60 mM **2** *into 8 mM* **1** *(first row) and 150 mM* **2** *into 10 mM* **3** *(last two rows) in MeOH at 25°C.*

obtained by nonlinear fitting the data with the one-site binding model. It is important to note that the thermodynamic parameters shown in **Table 4** are per mole of binding sites (n = 2), so the global association constant (which considers full site occupancy) is $(K_a)^2 \sim 10^5\ M^{-2}$ and the global values of ΔG, ΔH and ΔS are twice those reported, i.e.: $\sim -$ (28 ± 4) and -(34 ± 2) kJ/mol and $\sim -(20 \pm 20)$ kJ/mol K, respectively.

The sequential two-sites model was also employed to approach the experimental ITC data. The evolution of the residuals and the decrease of the standard deviation suggest that the sequential two-sites binding model better approximates the experimental titration data. This model assumes the existence of two distinct binding sites in **3** with 1:1 stoichiometry each (i.e., the population of type 1 sites and type 2 sites in the sample is identical). The fitting of the experimental data to the model therefore provides 4 thermodynamic parameters, 2 for the initial stage (occupancy of site 1) and 3 for the second stage (occupancy of site 2). Results are summarized in **Table 4**. Again, the global equilibrium constant is the product of the K_a obtained at each step $(K_{a1}xK_{a2}) \sim 10^5\ M^{-2}$, which is in full agreement with the result obtained using de 1:1 model. Similarly, the overall ΔG, ΔH and ΔS are the sum of the individual contribution to give: $\sim -$ (38 ± 6) and -(42 ± 4) kJ/mol and $\sim -(50 \pm 20)$ kJ/mol K, respectively. Within the experimental uncertainties, these values are similar to the ones obtained using the simplest one-site independent model. When the results determined with the sequential model for the **3–2** reaction are compared with those obtained for the **1–2** reaction, modeled with the one-site independent model, some interesting aspects emerge. For example, the first reaction step between **2** and **3** is spontaneous and its $\Delta G \sim -17$ kJ/mol is very similar to that obtained for the reaction of **2** with **1**, $\Delta G \sim -16$ kJ/mol. However, such coincidence occurs due to the compensation between ΔH and ΔS values (see **Table 4**), the entropic change for the first step of the reaction being for the reaction **3–2** positive. The results also suggest that the second reaction step is slightly less spontaneous, but is accompanied by a very large negative enthalpy and entropy change. Additional experimental data are required to better understand this system.

We should also mention that the two-sites independent model has also been tested but it has been discarded in the end as it contains the highest number of adjustable parameters (6 in total) and was giving unrealistic stoichiometries for the two sites.

5. Conclusion and perspectives

Isothermal calorimetry is a label free non-destructive technique that became over the years the method of choice for most binding studies in solution. Compared to other dedicated spectroscopic techniques like nuclear magnetic resonance, infrared spectrometry or fluorimetry, which require a series of experiments to extract site-specific stoichiometry and affinity information, ITC gives access - in a single experiment – to a complete thermodynamic picture of the overall interaction mechanism. The continuous development of the technique led to highly sensitive instruments capable to detect small amounts of samples. For example, isothermal micro- and nano-calorimetry can measure tiny heat changes and became an important method for the thermodynamics profile analysis of chemical and biochemical reactions, solvation and dissolution processes. The automation of the entire experimental process with the new generation of calorimeters optimizes not only the use of the calorimeter but also increases the number of samples that can be studied. In parallel, considerable efforts have been made to propose various binding models and softwares with powerful routines for the optimization of experimental conditions (notably in terms of concentrations).

Despite the widespread application of ITC to probe interactions between biomolecules, the technique is hardly ever used for the characterization of the complexation of small molecules, especially when the titration is performed in organic solvents. Herein, we presented few examples that support the high potential of ITC for the study of interactions between small molecules in methanol, acetonitrile and methanol/water mixture on a Nano ITC Low Volume device (TA Instruments), with an emphasis on both simple (1:1) and more complex (1:1 and 1:2) interaction mechanisms.

In addition to the binding studies at equilibrium, ITC provided promising results for the investigation of reaction kinetics, irreversible reactions, reactions under pressure, etc. The field of application of isothermal calorimetry is continuously expanding from pharmacology to life science, clinical medicine, environmental science, biotechnology, ecology, etc.

Acknowledgements

The authors thank Jacques Loubens at TA Instruments, Waters, France for his kind assistance with the ITCRun and data analysis NanoAnalyze softwares. We are grateful to Bernadette Tse Sum Bui for kindly providing the monomer (4-acrylamidophenyl)-aminomethaminium acetate. We thank the European Regional Development Fund and the Region of Picardy (CPER 2007-2020). L.D. acknowledges financial support from the European Commission (project NOSY for New Operational Sensing sYstems, Grant agreement ID 653839, H2020-EU.3.7), the Hauts-de-France Region, the European Regional Development Fund (ERDF) 2014/2020 and Idex Sorbonne Université Investissements d'Avenir (2019/2020 EMERGENCE program).

Conflict of interest

The authors declare no conflict of interest.

Author details

Raquel Gutiérrez-Climente[1†], Elise Prost[2†], Aude Cordin[2], Carlos Chesta[3] and Luminita Duma[4*]

1 University Montpellier, IBMM, CNRS, ENSCM, Montpellier, France

2 University of Technology of Compiègne, UPJV, UMR CNRS 7025, Enzyme and Cell Engineering, Research Centre Royallieu, Compiègne, France

3 National Universiy of Río Cuarto, IITEMA-CONICET, Río Cuarto, Argentina

4 Champagne-Ardenne University, CNRS, ICMR UMR 7312, Reims, France

*Address all correspondence to: luminita.duma@univ-reims.fr

† These authors contributed equally.

References

[1] Christensen JJ, Izatt RM, Hansen LD. New precision thermometric titration calorimeter. The Review of Scientific Instruments. 1965;**36**(6):779-783. DOI: 10.1063/1.1719702

[2] Hansen LD, Christensen JJ, Izatt RM. Entropy titration. A calorimetric method for the determination of ΔG, ΔH and ΔS. Chemical Communications. 1965;**70**(3):36-38. DOI: 10.1039/C1965 0000036

[3] Chaires JB. Calorimetry and thermodynamics in drug design. Annual Review of Biophysics. 2008;**37**:135-151. DOI: 10.1146/annurev. biophys.36.040306.132812

[4] Rizzuti B, Lan W, Santofimia-Castaño P, Zhou Z, Velázquez-Campoy A, Abián O, et al. Design of inhibitors of the intrinsically disordered protein nupr1: Balance between drug affinity and target function. Biomolecules. 2021;**11**(10):1-19. DOI: 10.3390/ biom11101453

[5] Hughes JP, Rees S, Kalindjian SB, Philpott KL. Principles of early drug discovery. British Journal of Pharmacology. 2011;**162**(6):1239-1249. DOI: 10.1111/j.1476-5381.2010.01127.x

[6] Linkuvienė V, Krainer G, Chen W-Y, Matulis D. Isothermal titration calorimetry for drug design: Precision of the enthalpy and binding constant measurements and comparison of the instruments. Analytical Biochemistry. 2016;**515**:61-64. DOI: 10.1016/j.ab. 2016.10.005

[7] Ladbury JE, Klebe G, Freire E. Adding calorimetric data to decision making in lead discovery: A hot tip. Nature Reviews. Drug Discovery. 2010;**9**(1): 23-27. DOI: 10.1038/nrd3054

[8] Velazquez-Campoy A, Luque I, Freire E. The use of isothermal titration calorimetry in drug design: Applications to high affinity binding and protonation/ deprotonation coupling. Netsu Sokutei. 2001;**28**(2):68-73. DOI: 10.11311/ jscta1974.28.68

[9] Freire E. Do enthalpy and entropy distinguish first in class from best in class? Drug Discovery Today. 2008; **13**(19):869-874. DOI: 10.1016/ j.drudis.2008.07.005

[10] Falconer RJ, Schuur B, Mittermaier AK. Applications of isothermal titration calorimetry in pure and applied research from 2016 to 2020. Journal of Molecular Recognition. 2021;**34**(10):e2901. DOI: 10.1002/jmr.2901

[11] Vander Meulen KA, Horowitz S, Trievel RC, Butcher SE. Measuring the kinetics of molecular association by isothermal titration calorimetry. Methods in Enzymology. 2016;**567**:181-213. DOI: 10.1016/bs.mie.2015.08.012

[12] Schmidtchen FP. Isothermal titration calorimetry in supramolecular chemistry. In: Anal. Methods Supramol. Chem. New Jersey, USA, Weinheim, Germany: John Wiley & Sons, Ltd, Wiley-VCH Verlag GmbH & Co. KGaA; 2006. pp. 55-78. DOI: 10.1002/ 9783527610273.ch3

[13] Biedermann F, Schneider H, Platz H. Experimental binding energies in supramolecular complexes. Chemical Reviews. 2016;**116**:5216-5300. DOI: 10.1021/acs.chemrev.5b00583

[14] Wang Y, Wang G, Moitessier N, Mittermaier AK. Enzyme kinetics by isothermal titration calorimetry: Allostery, inhibition, and dynamics.

Frontiers in Molecular Biosciences. 2020; 7:1-19. DOI: 10.3389/fmolb.2020.583826

[15] Di Trani JM, Moitessier N, Mittermaier AK. Measuring rapid time-scale reaction kinetics using isothermal titration calorimetry. Analytical Chemistry. 2017;**89**(13):7022-7030

[16] Di Trani JM, De Cesco S, O'Leary R, Plescia J, Do Nascimento CJ, Moitessier N, et al. Rapid measurement of inhibitor binding kinetics by isothermal titration calorimetry. Nature Communications. 2018;**9**(1):893. DOI: 10.1038/s41467-018-03263-3

[17] Braissant O, Wirz D, Göpfert B, Daniels AU. Use of isothermal microcalorimetry to monitor microbial activities. FEMS Microbiology Letters. 2010;**303**(1):1-8. DOI: 10.1111/j.1574-6968.2009.01819.x

[18] Medina S, Raviv M, Saadi I, Laor Y. Methodological aspects of microcalorimetry used to assess the dynamics of microbial activity during composting. Bioresource Technology. 2009;**100**(20):4814-4820. DOI: 10.1016/j.biortech.2009.05.015

[19] Gaisford S. Stability assessment of pharmaceuticals and biopharmaceuticals by isothermal calorimetry. Current Pharmaceutical Biotechnology. 2005; **6**(3):181-191. DOI: 10.2174/1389201054022913

[20] Kimura T, Matsushita T, Kamiyama T. Enthalpies of solution of aliphatic compounds in dimethyl sulfoxide. Thermochimica Acta. 2004; **416**(1–2):129-134. DOI: 10.1016/j.tca.2003.02.001

[21] Sugiura T, Ogawa H. Thermodynamic properties of solvation of aromatic compounds in cyclohexane, heptane, benzene, 1,4-dioxane, and chloroform at 298.15 K. The Journal of Chemical Thermodynamics. 2009; **41**(11):1297-1302. DOI: 10.1016/j.jct.2009.06.001

[22] Korolev VP, Antonova OA, Smirnova NL, Kustov AV. Thermal properties of tetraalkylammonium bromides in several solvents. Journal of Thermal Analysis and Calorimetry. 2011; **103**(1):401-407. DOI: 10.1007/s10973-009-0639-6

[23] Riveros DC, Martínez F, Vargas EF. Enthalpies of solution of methylcalix[4] resorcinarene in non-aqueous solvents as a function of concentration and temperature. Thermochimica Acta. 2012; **548**:13-16. DOI: 10.1016/j.tca.2012.08.022

[24] Alves N, Bai G, Bastos M. Enthalpies of solution of paracetamol and sodium diclofenac in phosphate buffer and in DMSO at 298.15 K. Thermochimica Acta. 2006;**441**(1):16-19. DOI: 10.1016/j.tca.2005.11.031

[25] Badelin VG, Mezhevoi IN. The thermochemical characteristics of solution of L-cysteine and L-asparagine in aqueous 1,4-dioxane and acetone. Russian Journal of Physical Chemistry A. 2009;**83**(7):1121-1124. DOI: 10.1134/S0036024409070139

[26] Mezhevoi IN, Badelin VG. Standard enthalpies of dissolution of L-alanine in the water solutions of glycerol, ethylene glycol, and 1,2-propylene glycol at 298.15 K. Russ. The Journal of Physical Chemistry. A. 2010;**84**(4):607-610. DOI: 10.1134/S0036024410040151

[27] Badelin VG, Smirnov VI. The dependence of the enthalpy of solution of L-methionine on the composition of water-alcohol binary solvents at 298.15 K. Russ. The Journal of Physical

Chemistry. A. 2010;**84**(7):1163-1168. DOI: 10.1134/S0036024410070150

[28] Smirnov VI, Badelin VG. Enthalpies of solution of glycylglycine and diglycylglycine in aqueous alcohols at 298.15 K. Thermochimica Acta. 2008;**471** (1–2):97-99. DOI: 10.1016/j.tca.2008.01.020

[29] Guseva GB, Antina EV, Berezin MB, V'yugin AI. Enthalpies of dissolution of cobalt(II) and copper(H) acetylacetonates in organic solvents. Russ. The Journal of Physical Chemistry. 2005;**79**(6):908-911

[30] Vandyshev VN, Ledenkov SF. Thermochemical study of complexation and solvation in β-alanine- Cu(NO3) 2-water-ethanol system: 1. Enthalpy characteristics of complexing copper(II) ion with β-alanine. Russian Journal of Inorganic Chemistry. 2011;**56**(3): 479-483. DOI: 10.1134/S0036023611030259

[31] Jóźwiak M, Madej-Kiełbik L. Effect of temperature on the process of complex formation crown ether 15C5 with Na+ in the (water + ethanol) mixture at temperatures from (293.15 to 308.15) K. Thermochimica Acta. 2014; **580**:13-19. DOI: 10.1016/j.tca.2014.01.019

[32] Usacheva TR, Pham Thi L, Terekhova IV, Kumeev RS, Sharnin VA. Application of isothermal titration calorimetry for evaluation of water-acetone and water-dimethylsulfoxide solvent influence on the molecular complex formation between 18-crown-6 and triglycine at 298.15 K. Journal of Thermal Analysis and Calorimetry. 2015; **121**(3):975-981. DOI: 10.1007/s10973-015-4630-0

[33] Bernaczek K, Mielańczyk A, Grzywna ZJ, Neugebauer D. Interactions between fluorescein isothiocyanate and star-shaped polymer carriers studied by isothermal titration calorimetry (ITC). Thermochimica Acta. 2016;**641**:8-13. DOI: 10.1016/j.tca.2016.08.007

[34] Arnaud A, Bouteiller L. Isothermal titration calorimetry of supramolecular polymers. Langmuir. 2004;**20**(16): 6858-6863. DOI: 10.1021/la049365d

[35] Wadsö I, Goldberg RN. Standards in isothermal microcalorimetry: (IUPAC technical report). Pure and Applied Chemistry. 2001;**73**(10):1625-1639. DOI: 10.1351/pac200173101625

[36] TA Instruments, Nano ITC. Available from: https://www.tainstruments.com/nano-itc/ [Accessed: January 15, 2022]

[37] Velázquez-Campoy A, Ohtaka H, Nezami A, Muzammil S, Freire E. Isothermal titration calorimetry. Current Protocols in Cell Biology. 2004;**23**:1-24. DOI: 10.1002/0471143030.cb1708s23

[38] Quin CF. Advanced ITC techniques: The reverse titration. 2013. http://www.tainstruments.com/pdf/literature/MCAPN-2013-01_Advanced_ITC_Techniques-The_Reverse_Titration.pdf

[39] Spuches AM, Kruszyna HG, Rich AM, Wilcox DE. Thermodynamics of the as(III)-thiol interaction: Arsenite and monomethylarsenite complexes with glutathione, dihydrolipoic acid, and other thiol ligands. Inorganic Chemistry. 2005;**44**(8):2964-2972. DOI: 10.1021/ic048694q

[40] Wiseman T, Williston S, Brandts JF, Lin LN. Rapid measurement of binding constants and heats of binding using a new titration calorimeter. Analytical Biochemistry. 1989;**179**(1):131-137. DOI: 10.1016/0003-2697(89)90213-3

[41] Myszka DG, Abdiche YN, Arisaka F, Byron O, Eisenstein E, Hensley P, et al. The ABRF-MIRG'02 study: Assembly state, thermodynamic, and kinetic analysis of an enzyme/inhibitor interaction. Journal of Biomolecular Techniques. 2003;**14**(4):247-269

[42] Bastos M, Velazquez-Campoy A. Isothermal titration calorimetry (ITC): A standard operating procedure (SOP). European Biophysics Journal. 2021;**50** (3–4):363-371. DOI: 10.1007/s00249-021-01509-5

[43] Campoy AV, Claro B, Abian O, Höring J, Bourlon L. A multi-laboratory benchmark study of isothermal titration calorimetry (ITC) using $-Ca^{2+}$ and $-Mg^{2+}$ binding to EDTA. European Biophysics Journal. 2021;**50**:429-451. DOI: 10.1007/s00249-021-01523-7

[44] Fisher HF, Singh N. Calorimetric determination of binding constants. Energy Biological Macromolecules. 1995; **259**:194-221. DOI: 10.1016/0076-6879 (95)59045-5

[45] Horn JR, Russell D, Lewis EA, Murphy KP. Van't Hoff and calorimetric enthalpies from isothermal titration calorimetry: Are there significant discrepancies? Biochemistry. 2001; **40**(6):1774-1778. DOI: 10.1021/bi002408e

[46] Mizoue LS, Tellinghuisen J. Calorimetric vs. van't Hoff binding enthalpies from isothermal titration calorimetry: Ba^{2+}-crown ether complexation. Biophysical Chemistry. 2004;**110**(1–2):15-24. DOI: 10.1016/j.bpc.2003.12.011

[47] Markova N, Hallén D. The development of a continuous isothermal titration calorimetric method for equilibrium studies. Analytical Biochemistry. 2004;**331**(1):77-88. DOI: 10.1016/j.ab.2004.03.022

[48] Degassing station. https://www.tainstruments.com/degassing-station/. [Accessed: February 4, 2022]

[49] Herrera I, Winnik MA. Differential binding models for direct and reverse isothermal titration calorimetry. The Journal of Physical Chemistry. A. 2016; **120**:2077-2086. DOI: 10.1021/acs.jpcb.5b09202

[50] Piñeiro Á, Muñoz E, Sabín J, Costas M, Bastos M, Velázquez-Campoy A, et al. AFFINImeter: A software to analyze molecular recognition processes from experimental data. Analytical Biochemistry. 2019;**577**:117-134. DOI: 10.1016/j.ab.2019.02.031

[51] Duvvuri H, Wheeler LC, Harms MJ. Pytc: Open-source python software for global analyses of isothermal titration calorimetry data. Biochemistry. 2018; **57**(18):2578-2583. DOI: 10.1021/acs.biochem.7b01264

[52] Gans P, Sabatini A, Vacca A. Simultaneous calculation of equilibrium constants and standard formation enthalpies from calorimetric data for systems with multiple equilibria in solution. Journal of Solution Chemistry. 2008;**37**(4):467-476. DOI: 10.1007/s10953-008-9246-6

[53] Le VH, Buscaglia R, Chaires JB, Lewis EA. Modeling complex equilibria in isothermal titration calorimetry experiments: Thermodynamic parameters estimation for a three-binding-site model. Analytical Biochemistry. 2013;**434**(2):233-241. DOI: 10.1016/j.ab.2012.11.030

[54] Zhao H, Piszczek G, Schuck P. SEDPHAT – A platform for global ITC analysis and global multi-method

analysis of molecular interactions. Methods. 2015;**76**:137-148. DOI: 10.1016/j.ymeth.2014.11.012

[55] Nicholls IA. Thermodynamic considerations for the design of and ligand recognition by molecularly imprinted polymers. Chemistry Letters. 1995;**24**(11):1035-1036. DOI: 10.1246/cl.1995.1035

[56] Scheuermann TH, Brautigam CA. High-precision, automated integration of multiple isothermal titration calorimetric thermograms: New features of NITPIC. Methods. 2015;**76**:87-98. DOI: 10.1016/j.ymeth.2014.11.024

[57] Dogadkin DN, Soboleva IV, Kuz'min MG. Formation enthalpy and entropy of exciplexes with variable extent of charge transfer in solvents of different polarity. High Energy Chemistry. 2001;**35**(4):251-257. DOI: 10.1023/A:1017636612185

[58] Solís C, Grosso V, Faggioli N, Cosa G, Romero M, Previtali C, et al. Estimation of the solvent reorganization energy and the absolute energy of solvation of charge-transfer states from their emission spectra. Photochemical & Photobiological Sciences. 2010;**9**(5):675-686. DOI: 10.1039/b9pp00190e

[59] Nestora S, Merlier F, Beyazit S, Prost E, Duma L, Baril B, et al. Plastic antibodies for cosmetics: Molecularly imprinted polymers scavenge precursors of malodors. Angewandte Chemie, International Edition. 2016;**55**:6252-6256. DOI: 10.1002/anie.201602076

[60] Panagiotopoulou M, Salinas Y, Beyazit S, Kunath S, Duma L, Prost E, et al. Molecularly imprinted polymer coated quantum dots for multiplexed cell targeting and imaging. Angewandte Chemie, International Edition. 2016; **55**(29):8244-8248. DOI: 10.1002/anie.201601122

[61] Job P. Formation and stability of inorganic complexes in solution. Annali di Chimica. 1928;**9**:113-203

[62] Renny JS, Tomasevich LL, Tallmadge EH, Collum DB. Method of continuous variations: Applications of job plots to the study of molecular associations in organometallic chemistry. Angewandte Chemie, International Edition. 2013;**52**(46):11998-12013. DOI: 10.1002/anie.201304157

[63] Goldberg RN, Kishore N, Lennen RM. Thermodynamic quantities for the ionization reactions of buffers. Journal of Physical and Chemical Reference Data. 2002;**31**(2):231-370. DOI: 10.1063/1.1416902

[64] Wang HM, Loganathan D, Linhardt RJ. Determination of the pK(a) of glucuronic acid and the carboxy groups of heparin by 13C-nuclear-magnetic-resonance spectroscopy. The Biochemical Journal. 1991;**278**(3):689-695. DOI: 10.1042/bj2780689

Chapter 2

A State of Art Review on Thermodynamics Performance Analysis in Pulse Detonation Combustor

Pinku Debnath and Krishna Murari Pandey

Abstract

Pulse detonation engines (PDEs) are most exciting for future propulsion generation. Detonation combustion in pulse detonation combustor is an energetic combustion process which is differs from other combustion process. The detonation wave propagation in detonation tube is a pulse setting combustion phenomena. Detonation combustion process is thousands times faster than deflagration combustion process. PDE utilizes several pulse of detonation wave to produce propulsive force. The potential applications of PDEs are drastically reduces the cost of orbit transfer vehicle system and flying mode applications. Of course it can be used as ground level applications also. Draw back are DDT in shortest possible time in the combustor. In this regards, worldwide researchers are focusing on scientific and technical issues related to improvement of PDC. The present chapter deals with review study on detonation combustion process, historical overview on chemical kinetics, calorimetric and entropy transport, energy and exergy analysis and factor effecting on deflagration to detonation transition with recommendable future research.

Keywords: pulse detonation combustor, CFD, detonation, deflagration, ejector

1. Introduction

The high speed engine concept was born in the early 1900s, which produced shaft work and designed to drive a variety of vehicles, including ships and locomotives, until further introduction of jet engine on 1930s. The history of pulse detonation engines concept can be traced back to German engineer Hoffmann, H. [1]. In 1941, they tested a prototype engine using acetylene-oxygen and benzene-oxygen mixtures. Earlier in between 1952 and 1956, Nicholls et al. [2] at the University of Michigan have independently come up with the idea of using intermittent detonation for propulsion system and built the first PDE, which utilized detonation of hydrogen-air mixture to produce thrust. When crude oil prices increased significantly in the mid-1980s Eidelman et al. reinitiated research on PDE to overcome these scarcities. Krzycki [3] experimentally studied the propane-air pulse detonation engine which is operating at 25–50 Hz at naval ordnance test station at China Lake, California on 1962. From this analysis it was observed that minimum thrust is produce at minimum operating frequency. Lynch et al. [4] performed CFD studies on PDREs and air breathing PDEs on 1990. Their analyses forecasted that PDEs would be incorporated with space transport vehicles by the early 2000s. Earlier starting in 1990s, experimental study of single and multi-pulse detonation engine combustor were conducted by Bussing et al. [5] at Adroit systems Inc., a company that was bought up by Pratt and Whitney on 2001. They tested pulse detonation engine using different fuels including hydrogen and ethylene. Most of the PDE research centers are found in Canada, US, Russia, China. There are very less number of PDE research centre in India, so researcher are focusing on pulse detonation research area as it is excited for future propulsion technology [6].

Propulsion applications of detonation can be classified into three categories: standing detonation, pulse detonation and rotating detonation [7–9]. The basic pulse detonation engine has a very simple structure. It consists of a constant area tube. The deflagration to detonation transition is controlled by supplying fuel and oxidizer in detonation tube. The ignition system and nozzle are used for accelerating the flow, which is to be used for propulsion. A practical pulse detonation engine may also have one or more devices to bring about deflagration to detonation transition such device are Shchelkin spiral and blockage [10]. The PDE consists of two or more combustion chamber, which is joined to common plenum chamber. The conditions are applied for accelerating the flow before entering the detonation tube with different nozzle. The can-annular four chambered PDE is can illustrated for propulsion system. In multi-chambered design, each chamber can be at different stage in the cycle, thus creating a smoother flow through the nozzle [11]. The ejector enhances the deflagration to detonation transition in detonation tube with an array setting in exit section of pulse detonation rocket engine. Another feature of the ejector design is that the detonation waves from the combustors can be used to enhance the propulsive performance, which provides additional thrust enhancement [12].

Detonation is a supersonic mode of combustion process. In combustion process detonation waves are much more energetic process than conventional deflagration combustion and it produces a very strong wave coupled with a chemical reaction zone, propagating at supersonic speed. A detonation wave compresses combustion mixture, increasing the combustion product pressure, density of species mass fraction. It is a subsonic combustion process and fuel air reaction propagates at relatively low speed and reasonably low pressure from a trailing reaction zone. The propagation of deflagration mode of combustion consists of diffusion of unburned gases ahead of flame front and burnt gases behind the combustion flame. Deflagration produces small decrease in pressure and can be modeled as a constant pressure process [13]. One of the primary attributes deflagration flame travels at a speed, which is significantly lower than that the speed of sound ($Ma < 1$). So it can be identify by subsonic combustion process. Detonation combustion is a constant volume combustion process. The strength of leading shock depends on the detonation wave propagation velocity. A simple planar model for the supersonic detonation shock wave is used for Chapman-Jouguet detonation model analysis [14]. This is a rapid exothermic reaction and instantaneously changes the local pressure and temperature. The ignition of fuel-air mixture can produce deflagration flame and later on transition to detonation wave. The different combustor geometry can accelerate the deflagration flame and transition to detonation wave. Several researchers have been studied on PDE with research gape and scope of future research work [15–17]. The applications of RDE chamber are jet engines, such as turbojet or gas turbine, ramjet or rocket. Continuous detonation wave engine is used for supersonic and hypersonic propulsion applications. The framework of French Research and Development and scientific research also consider these for space applications. However incoming reacting air-mixture is greater than the C-J velocity of fuel-air mixture. Such engine is scramjet engine with an oblique detonation wave at inlet to combustor called the oblique detonation wave engine [18]. Chapman [19] explained on 1889 that the minimum speed of burnt gas is equal to speed of sound in gas mixture. Later on Jouguet on 1905 [20] applied Hugoniot's method to explain the detonation velocity. The explosive mixture can get supported with two modes of combustion. When the flame propagates at slow velocity relative to unburnt gases, it is define as deflagration mode. In detonation mode wave propagates at about 2000 m/s accompanied by an overpressure rise is near about 20 bars [21–23]. They independently developed the basic thermodynamic model behind detonation. Principle operation of standing detonation engine is relatively simple. Fuel is injected into supersonic flow and detonation wave is stabilized inside the engine by wedge or other means and products are expanding inside nozzle. The combustion wave velocity can be propagates at higher the C-J detonation velocity within the Mach number of 5. The principle of rotating detonation engine (RDE) is based on the formation of detonation in a disk type combustion chamber. The shape of combustion chamber is toroidal or ring-like shape [24, 25]. The detonation wave parameters are depending on critical detonation tube diameter and minimum detonation tube diameter. The minimum and critical diameters are important parameters for evaluation of performance of PDE. The detonation will successfully propagate in a tube when the diameter must be larger than $\lambda/3$, where λ is the cell size. For square and rectangular ducts, the width and height of the duct must be larger than λ [26]. A review of the gas dynamics and chemistry of real detonation is discovered by Fickett and Davis [27]. They found out initiation of detonation wave, which follows by a series of percentage of fuel-oxidizer mixture in combustion chamber. The detonation wave in a confined tube causes the reaction of fuel-air mixtures, which creates turbulence; as a result "an explosion in an explosion" is takes placed. The two strong shock waves are created in the opposite direction, the forward shock waves are known as retonation. A self-propagating C-J detonation wave is formed at steady state retonation process. The pre-detonation wave velocity is 1000 m/s while the characteristic C-J detonation speed is over 2000 m/s. A large explosion occurs at onset of detonation, resulting in an over-driven detonation wave that decays to the C-J velocity. The wall roughness controls the wave propagation by inducing large-amplitude unsteady and turbulent flow, complex wave interaction processes and high temperature behind shock reflections. These effects represent ways that the flow can generate large-scale turbulence for flame folding and large temperature fluctuations causing detonation initiation [28].

2. Review on thermodynamics cycle analysis

A pulse detonation engine uses repetitive cycle of detonation waves to combust fuel-oxidizer mixture for producing thrust. PDE operates by propagating detonation wave through a tube filled with a combustible mixture and generates propulsive thrust. This process results are near about constant volume combustion process, which produces high pressures from the leading shock wave. Pulse detonation engine consists of valve less combustor with straight tube, which is closed at one end and open at other end. The pulse detonation engine combustion cycle consist of four basic thermodynamics process. The first process is filling time (t_{fill}) of fuel-air mixture for detonation combustion, which is estimated as length of the tube over filling velocity. The second one is detonation combustion. This process takes place within fraction of millisecond. As soon as the detonation wave reaches to the closed end region the pressure and velocity decrease from initial position to exit end region. Fully developed detonation wave travels with the magnitude of Chapman-Jouguet speed. This C-J speed of reacting fuel-air mixture varies between 1400 m/s and 1800 m/s. The detonation time of the wave (t_c) is therefore similarly estimated by the length of the tube over the C-J wave velocity. The time required for blow down (t_b) stage can be estimated by the length of the tube to the rarefaction velocity. At last in the purging process the tube is scavenged off hot detonation products by using fresh air. Purging process is necessary to prevent auto ignition of the fresh fuel-oxidizer mixture. The time taken for purging the tube with the fresh air (t_{purge}) is the length of the tube over the purging velocity [29]. So total sum of the time for all the four stages are as follows:

$$\mathrm{T} = t_{fill} + t_c + t_b + t_{purge} \tag{1}$$

The PDE can run by any fuel, liquid or gaseous, like natural gas, propane, bio-gas, hydrogen, kerosene, jet fuels and octane etc. From an engineering stand point fuel can be selected for based on heating value, detonability, ignition time, energy release, adiabatic flame temperature and sensivity with air [30]. Povinelli and Yungster [31] studied the thermodynamic cycle of hydrogen-air mixture at static conditions in pulse detonation combustor. The specific thrust, fuel consumptions and impulse of detonation combustion are analyzed by using CFD analysis with finite rate chemistry. Alam et al. [32] studied on Brayton, Humphery and ideal thermodynamics cycle analysis in pulse detonation combustor. They found Humphery cycle efficiency can be increases with higher value of compression ratio. The thermodynamics cycle efficiency of air breathing pulse detonation engine is studied by Wu et al. [33]. They found that chocked convergent-divergent nozzle is required to improve the efficiency. Vutthivithayarak et al. [34] discussed the Humphrey and F. J. (Fickett-Jacobs) cycles in PDE. These cycles are illustrated with hydrogen-air combustion for generic heat release.

3. Review on chemical kinetics and entropy transport

The two-step chemical kinetics model of detonation combustion has been studied by Fomin [35]. This kinetics model has been used for stoichiometric, lean and rich mixture for combustion. This model is also followed by Le Chatelier's principle and 2nd law of thermodynamics. The pulse detonation combustor has lower entropy change and self-pressure gain compared to isobaric combustion process for same operating conditions [36]. Mehdi Safari et al. [37] studied on entropy generation with species transport equation for detonation combustion by large eddy simulation. Detonation initiation in hydrogen-air depends on mixture sensivity and geometrical parameters. Qi et al. [38] investigated the thermodynamics characteristics of methane-air detonation in pressure gain combustor. They compared the entropy change in detonation combustion process with gas turbine cycle. They found that cycle efficiency enhance rate up to 11.89%. Lu et al. [39] studied on DDT in a channels with obstacles using chemical diffusive model (CDM) integrated with reactive

Navier stokes equation. They found that CDM reduces the ignition time of detonation wave. Wu et al. [40] studied on atomization of liquid fuel detonation combustion. They found that nozzle can effectively atomize fuel-air mixtures under high pressure condition. Maciel and Marques [41] studied on hydrogen fuelled single cycle pulse detonation engine in Ansys Fluent. When OH^* kinetics added to the reaction set, they found cellular structure of detonation wave front in reaction zone. Ivanov et al. [42] studied on hydrogen-oxygen flame acceleration and transition from DDT in a channel using reactive Navier-Stokes equations. They found that steady detonation wave front is form in wider detonation channels of 10 mm and closed to C-J detonation propagation speed. Srihari et al. [43] studied on stoichiometric ethylene-air mixture of detonation combustion with one-step overall reaction model. They found that chemical reaction models have capable to predict the detonation wave velocity with reasonable accuracy.

4. Review on energy and exergy analysis

Ma et al. [44] studied on temporal variation of activation energy release rate of iso-octane vapor-air mixture in an obstacle-filled detonation tube. Their result shows that the activation energy influences the flame propagation parameters and deflagration-to-detonation transition process. Hutchins and Metghalchi [45] studied on exergy analysis of pulse detonation engine. They found that during deflagration to detonation transition period exergy loss is more. Bellini and Lu [46] studied on exergy analysis of fuel-air mixture at high frequency source within the detonation chamber. They found that combustion product accelerates inside the combustor in presence of Shchelkin spiral. The exergy analysis of pulse detonation power device designed for power production using gaseous fuel methane (CH_4) and propane (C_3H_8) is analysis by Bellini and Lu [47]. The exergetic efficiency was analyzed for different cycle frequency corresponding to detonation tube length. Rouboa et al. [48] studied on exergy loss of hydrogen-air detonation during shock. They also observed exergy destruction increase with augmentation of hydrogen concentration in reacting mixture. Petela [49] studied on exergy analysis of gaseous fuel-air detonation. They observed that exergy gives a quantitative theoretical useful work that is obtained from different energy form combustion process and it is a function of system and environment. Som and Datta [50] and Som and Sharma [51] studied on theoretical model of energy and exergy balance in a spray combustion process. They found that exergy destruction in this combustion process can be reduced through proper control of chemical reactions.

5. Results from CFD simulation and calorimetric analysis

The numerical investigations have been done in Ansys fluent platform. The **Figure 1** shows that the ejector effects on unsteady detonation combustion wave phenomena in pulse detonation combustor. The time dependent detonation wave contour plots clearly shows that 0.033 seconds is required to reach the fully developed detonation wave [52]. They also found that ejector plays the vital role for vortex formation of reacting mixture in PDE combustor. They also observed that leading vortex rings are found in shrouded ejector taper angle of +4°. The **Figure 2**

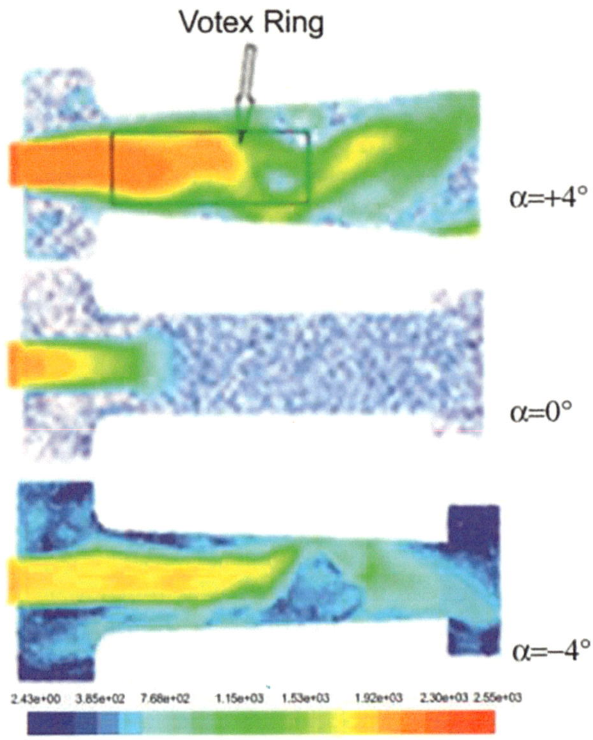

Figure 1.
Effect of shrouded ejectors on vortex ring formation of detonation wave [53].

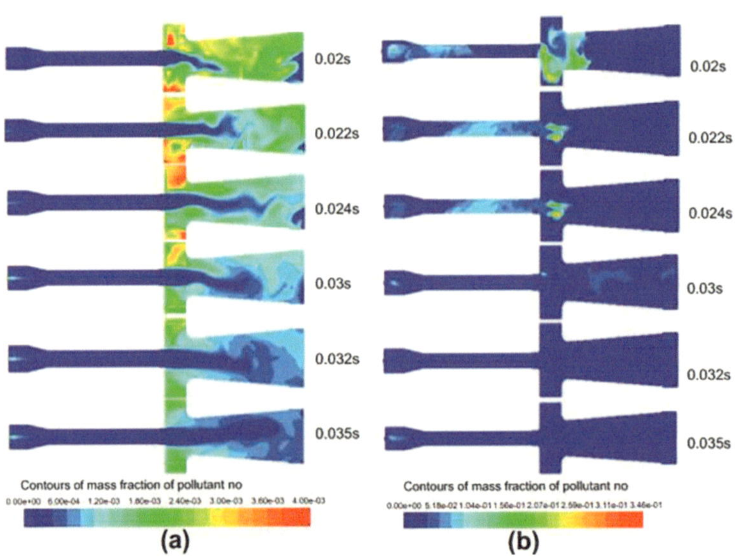

Figure 2.
The mass fraction contour analysis of NO_x pollutant number of (a) hydrogen-air and (b) kerosene-air combustion [53].

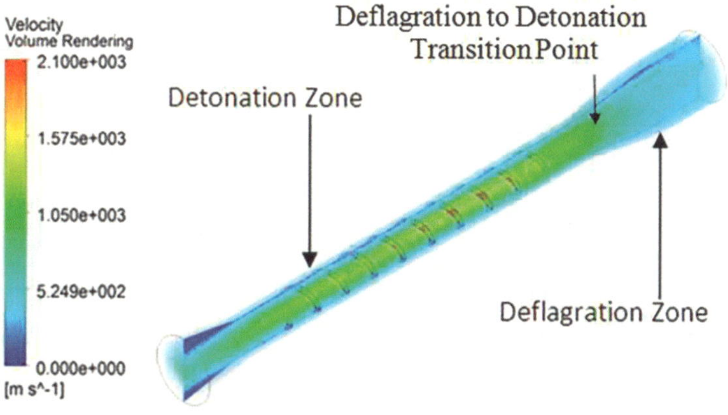

Figure 3.
Deflagration and detonation zone define by C-J velocity for exergy analysis [54].

shows the mass fraction contour analysis of NO_x pollutant number of hydrogen-air and kerosene-air detonation [53]. Lesser the fuel mass fraction higher the exergetic efficiency was found in pulse detonation combustor. The **Figure 3** shows the deflagration and detonation control volume for exergy analysis [54]. Alam et al. [55, 56] numerically studied on hydrogen-air detonation in pulse detonation combustor. Later on they also studied detonation combustion using alternative fuels, i.e. octane (C_8H_{18}), hexane (C_6H_{14}), pentane (C_5H_{12})-air combustion in PDE combustor. They observed combustion efficiency of pentane-air mixture is higher than that of other fuels. Alam et al. [57] studied the combustion wave propagation in obstructed detonation tube. Their simulation results were carried out for stoichiometric mixture of kerosene-air and butane-air mixture at atmospheric conditions. They found that mixing of butane-air combustion process is better than kerosene-air mixture. Furthermore, the stoichiometric ethane-air (C_2H_6-air) and ethylene-air (C_2H_4-air) fuel mixture at atmospheric pressure conditions has been studied by Alam et al. [58]. The effect of blockage ratio of 0.4, 0.5, 0.6 and 0.7 in channel for detonation wave acceleration are shown in **Figure 4**. The contour plot analysis shows the shock wave initiation and propagation time period in detonation tube is reduced by smaller blockage ratio of 0.5 [59]. Tripathi et al. [60] computationally studied on effect of obstacle on flame propagation velocity. Alam et al. [61] studies on flame acceleration in pulse detonation engine with changing the obstacle clearance. They found that combustor pressure is reduced as increase the obstacle clearance. P. Debnath and Pandey [62] studied on deflagration to detonation transition in PDE combustor with Schelkin spiral effect inside the detonation tube. They found that Schelkin spiral accelerate the flame propagation. Alam et al. [63] numerically studied on flame propagation in obstructed pulse detonation combustor with hydrogen-air mixture. They found that performance is increase up to 4.46% and this value increase for $\phi = 1.3$. Debnath and Pandey [64] studied on effect of different nozzle on flame acceleration and they found that divergent nozzle has more effect on flame acceleration. The **Figure 5** shows the comparison of thrust power for PDE combustor with several nozzle. Chourasia et al. [65] studied on progress and motivation of research in pulse detonation combustor. Xudong Zhang et al. [66] studied on critical mode

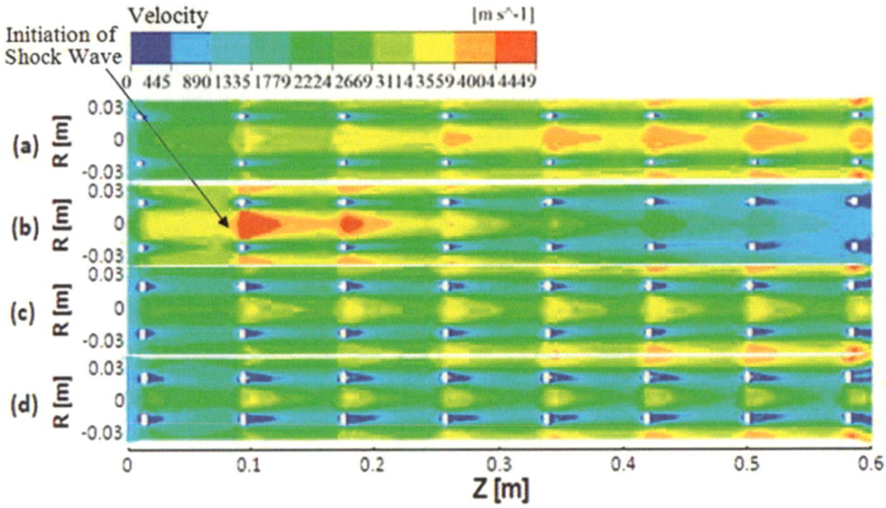

Figure 4.
Effect of blockage ratio on detonation wave propagation [59].

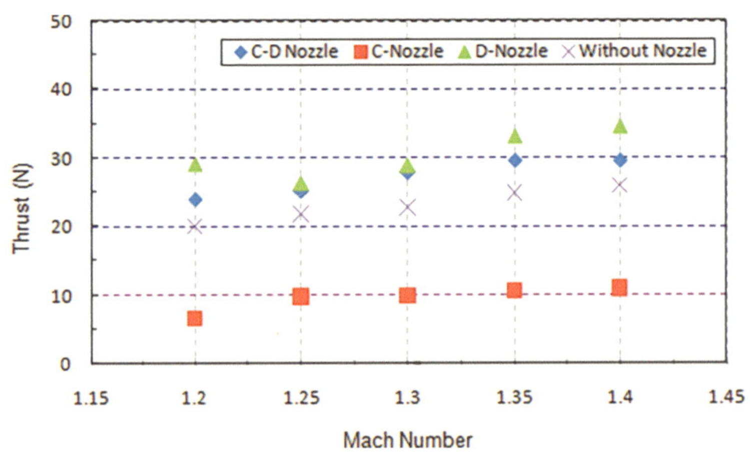

Figure 5.
The propulsive thrust variation for PDE combustor with C-D nozzle, C-nozzle, D-nozzle and without nozzle at different Mach number [64].

of gaseous methane-air detonation propagation in an annular tube based on reactive Navier Stokes Equations. They found that trajectories of triple point of the shock wave cell structures are petal pattern. Wang et al. [67] studied on effect of oxygen concentration on propane-air detonation in pulse detonation engine with straight nozzle, convergent nozzle, and convergent-divergent (CD) nozzle. Their results indicate that for the PDE with straight nozzle requires the shortest possible time for reacting gas burnt with high-temperature in detonation tube. Jishnu Chandran and Salih [68] studied on development of a benchmark solution in compressible liquid flows for shock tube problems. The compressibility effects in liquid water have been studied using the high-accuracy modified NASG equation of state. Arjun Singh et al. [69] studied on thermodynamic parameters for the formation of activation energy and self-acceleration for thermal explosion from critical temperature. They found that the thermal stability has been significantly reduced in presence of hydroxyl-terminated poly butadiene. Dong et al. [70] studied on correlations among detonation velocity, thermal stability, heat of combustion and decomposition kinetics of nitric esters. They found that oxygen coefficient plays positive role on decomposition of heat release efficiency of detonation combustion.

6. Concluding remarks

The above literature survey represents that there is more research is needed in pulse detonation combustor for shortest possible pulse time of deflagration to detonation transition. The future proposed research can be analyzed by changing the design of PDE combustor and operating conditions. The series of numerical simulations and optimization can be performed desire research objectives of pulse detonation engine. From the CFD and calorimetric analysis the smaller blockage ratio of 0.5 is found better to reduce detonation wave run up distance. The ejector enhance the shortest possible time of 0.033 s, which is required for fully developed detonation wave. More possible pulse time can be reduced by ejector geometry modification. Lesser the hydrogen fuel mass fraction of 0.25 higher the exergetic efficiency of 67.55% is obtained from detonation combustion process. Once the computational model is validated, further simulation can be carried out with accuracy. There are several detonation tube geometry is steal in debate for acoustics atomization and evaporative characteristics of liquid fuel detonation wave.

Author details

Pinku Debnath[1]* and Krishna Murari Pandey[2]

1 Department of Mechanical Engineering, National Institute of Technology Agartala, Agartala, Tripura, India

2 Department of Mechanical Engineering, National Institute of Technology Silchar, Silchar, Assam, India

*Address all correspondence to: er.pinkunits@yahoo.com

References

[1] Hoffmann H. Reaction-propulsion produced by intermittent detonative combustion. German Research Institute for Gliding, Report ATI-52365. 1940

[2] Nicholls JA, Wilkinson HR, Morrison RB. Intermittent detonation as a thrust-producing mechanism. Journal of Jet Propulsion. 2012;**27**(5):534-541

[3] Krzycki LJ. "Performance characteristics of an intermittent detonation device," NAVWEPS, Report 7655, Naval ordnance test station. China Lake, California: 1962

[4] Lynch ED, Edelman R, Palaniswamy S. Computational fluid dynamic analysis of the pulse detonation wave engine concept. In: 32nd Aerospace Sciences Meeting & Exhibit; January 10-13; Reno, NV. 2012

[5] Bussing T, Bratkovich TE, Hinkey JB. Practical implementation of pulse detonation engines, AIAA-1997-2748. In: 33rd AIAA/ASME/SAE/ASEE Joint Propulsion Conference and Exhibit; July 6-9; Seattle, WA. 2012

[6] Lu FK. Pulse Detonation Propulsion Systems Introduction. Texas: The University of Texas Arlington; 2007

[7] Bussing T, Pappas G. An introduction to pulse detonation engines. In: 32nd Aerospace Sciences Meeting and Exhibit; Reno, NV. 2012

[8] Eidelman S, Grossmann W, Lottati I. Review of propulsion applications and numerical simulations of the pulsed detonation engine concept. Journal of Propulsion and Power. 1991;7(6):857-865

[9] Li JL, Fan W, Qiu H, Yan C, Wang YQ. Preliminary study of a pulse detonation wave engine. Aerospace Science and Technology. 2010;**14**:161-167

[10] Panicker PK. The development and testing of pulsed detonation engine ground demonstrators [Ph.D. thesis]. The University of Texas at Arlington; 2008

[11] Oates G. Aerothermodynamics of Gas Turbine and Rocket Propulsion. 3rd ed. American Institute of Aeronautics & Astronautics (AIAA). 1997

[12] Munipalli R, Shankar V, Wilson DR, Lu FK. Preliminary design of a pulse detonation based combined cycle engine. In: 15th International Symposium on Air Breathing Engines; 2-7 September; Bangalore, India. 2001

[13] Kuo KK. Principles of Combustion. 2nd ed. John Wiley & Sons; 2005. ISBN 10: 9780471046899

[14] Nekkanti K. Analysis of thrust development in a pulse detonation engine [M.S. thesis]. The University of Texas at Arlington. Libraries Research Commons; 2010

[15] Pandey KM, Debnath P. Reviews on recent advances in pulse detonation engines. Journal of Combustion. 2016;**2016**:1-16

[16] Debnath P, Pandey KM. Performance investigation on single phase pulse detonation engine using computational fluid dynamics. In: Proceedings of the ASME 2013 International Mechanical Engineering Congress & Exposition, IMECE2013; November 15-21; San Diego, CA. 2013

[17] Pandey KM, Kumar J. CFD analysis of pulse detonation engine: A review. Journal of Material Science and Mechanical Engineering (JMSME). 2016;**3**(2):111-116

[18] Kailasanath K. Application of detonations to propulsion: A review. In: 37th AIAA Aerospace Sciences Meeting and Exhibit; January 11-14; Reno, NV. 1999

[19] Rankine WJM. On the thermodynamic theory of finite longitudinal disturbance. Philosophical Transactions of the Royal Society of London. 1870;**160**:277-288

[20] Lee JHS. Dynamic parameters of gaseous detonations. Annual Review of Fluid Mechanics. 1984;**16**:311-336

[21] Fickett W, Davis WC. Detonation Theory and Experiment. India: Dover Publications Inc; 1979. ISBN 10: 0486414566

[22] Taylor G. The dynamics of the combustion products behind plane and spherical detonation fronts in explosives. Proceedings of the Royal Society of London A. 1950;**200**(1061):235-247

[23] Wintenberger E. Application of steady and unsteady detonation waves to propulsion [Ph.D. dissertation]. Pasadena, CA: California Institute of Technology; 2004

[24] Kindracki J, Wolanski P, Gut Z. Experimental research on the rotating detonation in gaseous fuels-oxygen mixture. Shock Waves. 2011;**21**(2):75-84

[25] Hugoniot PH. Memoire sur la propagation du movement dans les corps et plus specialement dans les gaz parfaits premiere partie. Journal de l'Ecole Polytechnique. 1887;**57**:3-97

[26] Lee JHS. The Detonation Phenomenon. Cambridge University Press; 2014

[27] Fickett W, Davis WC. Detonation: Theory and Experiment. Dover Publications; 2010

[28] Oran ES, Gamezo VN. Origins of the deflagration-to-detonation transition in gas-phase combustion. Combustion and Flame. 2007;**148**:4-47

[29] Mattingly JD. Elements of Propulsion: Gas Turbines and Rockets. Reston, VA: AIAA; 2006

[30] Alhussan K, Assad M, Penazkov O. Analysis of the actual thermodynamic cycle of the detonation engine. Applied Thermal Engineering. 2016;**107**:339-344

[31] Povinelli LA, Yungster S. Thermodynamic Cycle and CFD Analyses for Hydrogen Fueled Air-breathing Pulsed Detonation Engines. American Institute of Aeronautics and Astronautics. National Aeronautics and Space Administration, Glenn Research Center; 2002

[32] Alam N, Sharma KK, Pandey KM. Thermodynamic performance of pulse detonation engine: A Technical Report". In: Proceedings of International Conference on Sustainable Computing in Science, Technology and Management (SUSCOM); February 26-28, 2019; Amity University Rajasthan, Jaipur, India. 2019. DOI: 10.2139/ssrn.3355295

[33] Wu Y, Ma F, Yang V. System performance and thermodynamics cycle analysis of air breathing pulse detonation engine. Journal of Propulsion and Power. 2003;**18**(4):556-567

[34] Vutthivithayarak R, Braun EM, Lu FK. On thermodynamics cycle for detonation engine. In: 28th International Symposium on shock waves. Berlin, Heidelberg: Springer; 2012

[35] Fomin PA. Reduced chemical kinetic model of detonation combustion of one and multi fuel gaseous mixtures with air. AIP Conference Proceedings. 2018;**1939**:020012

[36] Wintenberger E, Sheperd JE. Thermodynamic cycle of analysis for propagating detonations. Journal of Propulsion and Power. 2006;**22**:694-698

[37] Mehdi S, Reza M, Sheikhi H, et al. Entropy transport equation in large Eddy simulation for exergy analysis of turbulent combustion systems. Entropy. 2010;**12**:434-444. DOI: 10.3390/e1230434

[38] Qi L, Wang Z, Zhao N, Dai Y, Zheng H, Meng Q. Investigation of the pressure gain characteristics and cycle performance in gas turbines based on Interstage bleeding rotating detonation combustion. Entropy. 2019;**21**:265

[39] Lu X, Kaplan CR, Oran ES. Predictions of flame acceleration, transition to detonation and detonation propagation using the chemical-diffusive model. Combustion and Flame. 2022;**235**:111705

[40] Wu Y, Han Q, Yang G. Effects of an acoustic atomizer upon liquid-fueled detonation initiations in a detonation tube. Experimental Thermal and Fluid Science. 2019;**109**:109863

[41] Maciel EC, Marques CST. 2-D simulation with OH* kinetics of a single-cycle pulse detonation engine. Journal of Applied Fluid Mechanics. 2019;**12**(4):1249-1263

[42] Ivanov MF, Kiverin AD, Liberman MA. Flame acceleration and DDT of hydrogen-oxygen gaseous mixtures in channels with no-slip walls. International Journal of Hydrogen Energy. 2011;**36**(13):7714-7727

[43] Srihari P, Mallesh MA, Sai Krishna G, Charyulu BVN, Reddy DN. Numerical study of pulse detonation engine with one-step overall reaction model. Defence Science Journal. 2015;**65**(4):265-271

[44] Hu M, Xia Z, Gao W, Zhuo C, Wang D. Numerical simulation of the deflagration to detonation transition of iso-octane vapor in an obstacle-filled tube. International Journal of Spray and Combustion Dynamics. 2018;**10**(3):244-259

[45] Hutchins TE, Metghalchi M. Energy and exergy analysis of the pulse detonation engine. Journal of Engineering for Gas Turbines and Power. 2003;**125**(4):1075-1080

[46] Bellini R, Lu FK. Exergy analysis of a pulse detonation power device. In: Proceedings of the 10th Brazilian Congress of Thermal sciences and Engineering-ENCIT 2004, Braz. Soc. of Mechanical Sciences and Engineering-ABCM; November 29–December 03; Rio de Janerio, Brazil. 2004

[47] Bellini R, Frank KL. Exergy analysis of a hybrid pulse detonation power device. Journal of Propulsion and Power. 2010;**26**(4):875-878. DOI: 10.2514/1.4414

[48] Rouboa A, Silva V, Couto N. Exergy analysis in hydrogen-air detonation. Journal of Applied Mathematics. 2012;**2012**:16

[49] Petela R. Application of exergy analysis to the hydrodynamic theory of detonation in gases. Fuel Processing Technology. 2000;**67**:131-145

[50] Som SK, Datta A. Thermodynamic irreversibilities and exergy balance in combustion processes. Progress in Energy and Combustion Science. 2008;**34**:351-376

[51] Som SK, Sharma NY. Energy and exergy balance in the process of spray combustion in a gas turbine combustor. ASME Journal of Heat Transfer. 2002;**124**:828-836

[52] Debnath P, Pandey KM. Numerical investigation of detonation combustion wave in pulse detonation combustor with ejector. Journal of Applied Fluid Mechanics. 2017;**10**(2):725-733

[53] Debnath P, Pandey KM. Numerical analysis of detonation combustion wave in pulse detonation combustor with modified ejector with gaseous and liquid fuel mixture. Journal of Thermal Analysis and Calorimetry. 2021;**145**:3243-3254

[54] Debnath P, Pandey KM. Exergetic efficiency analysis of hydrogen-air detonation in pulse detonation combustor using computational fluid dynamics. International Journal of Spray and Combustion Dynamics. 2016;**9**(1):44-54

[55] Alam N, Pandey KM, Sharma KK. Combustion characteristics of hydrogen-air mixture in pulse detonation engines. Journal of Mechanical Science and Technology. 2019;**33**(5):1-7

[56] Alam N, Sharma KK, Pandey KM. Effects of various composition of fuel-air mixture on performance of pulse detonation engine. Combustion, Explosion, and Shock Waves. 2019;**55**:708-717

[57] Alam N, Pandey KM, Sharma KK. Numerical investigation of combustion wave propagation in obstructed channel of pulse detonation engine using kerosene and butane fuels. Journal of Applied Fluid Mechanics. 2019;**12**(3):883-890

[58] Alam N, Sharma KK, Pandey KM. Numerical investigation of combustion phenomena in pulse detonation engine with different fuels. AIP Conference Proceedings. 2018;**1966**:020015

[59] Debnath P, Pandey KM. Effect of blockage ratio on detonation flame acceleration in pulse detonation combustor using CFD. Applied Mechanics and Materials. 2014;**656**:64-71. DOI: 10.4028/www.scientific.net/AMM. 656.64

[60] Tripathi S, Pandey KM, Randive P. Computational study on effect of obstacles in pulse detonation engine. International Journal of Engineering & Technology. 2018;**7**:113-117

[61] Alam N, Sharma KK, Pandey KM. Numerical investigation of flame propagation in pulse detonation engine with variation of obstacle clearance. Journal of Thermal Analysis and Calorimetry. 2021;**140**:2485-2495

[62] Debnath P, Pandey KM. Computational study of deflagration to detonation transition in pulse detonation engine using Shchelkin spiral. Applied Mechanics and Materials. 2015;**772**:136-140

[63] Alam N, Sharma KK, Pandey KM. Numerical investigation of fame propagation and performance of obstructed pulse detonation engine with variation of hydrogen and air. Journal of the Brazilian Society of Mechanical Sciences and Engineering. 2019;**41**:502

[64] Debnath P, Pandey KM. Numerical investigation of detonation combustion wave propagation in pulse detonation combustor with nozzle. Advances in Aircraft and Spacecraft Science. 2020;**7**(3):187-202

[65] Chourasia A, Verma KA, Pandey KM. Review of computational work in pulse detonation engines. International Journal of Innovative Technology and Exploring Engineering. 2019;**8**:398-401

[66] Zhang X, Gui M, Pan Z, Zhang H. Numerical investigation on the rotating detonation critical mode for a methane-air mixture in an annular tube using

reactive Navier-Stokes equations. Journal of Thermal Analysis and Calorimetry. 2021;**144**:2285-2293

[67] Wang Z, Wei L, Qin W, Liang Z, Zhang K. Oxygen concentration distribution in a pulse detonation engine with nozzle-ejector combinational structures. Proceeding of IMechE Part G: Journal of Aerospace Engineering. 2021;**235**:1-10

[68] Jishnu Chandran R, Salih A. Development of a benchmark solution in compressible liquid flows: Analytical solution to the water shock tube problem. Journal of Thermal Analysis and Calorimetry. 2021. [Published online: 08 June 2021]

[69] Singh A, Soni PK, Sarkar C, Mukherjee N. Thermal reactivity of aluminized polymer-bonded explosives based on non-isothermal thermogravimetry and calorimetry measurements. Journal of Thermal Analysis and Calorimetry. 2019;**136**:1021-1035

[70] Dong J, Yan Q-L, Liu P-J, He W, Qi X-F, Zeman S. The correlations among detonation velocity, heat of combustion, thermal stability and decomposition kinetics of nitric esters. Journal of Thermal Analysis and Calorimetry. 2018;**131**:1391-1403

Chapter 3

Assessment of the Heat Capacity by Thermodynamic Approach Based on Density Functional Theory Calculations

Viorel Chihaia, Valentin Alexiev and Hasan S. AlMatrouk

Abstract

The theoretical aspects of the thermodynamic calculation of the Gibbs energy and heat capacity of a crystalline system within the frame of the Density Functional Theory (DFT) are introduced in the present chapter. Various approximations of phonon motion (harmonic, quasiharmonic, and anharmonic) and their effects on the thermodynamic properties are discussed. The theoretical basis of the thermodynamic approach of the heat capacity of crystals for given thermodynamic conditions is presented, having as example six polymorphs of the magnesium hydrides.

Keywords: density functional theory, phonons, thermodynamic calculations, Gibbs free energy, heat capacity, magnesium hydrides

1. Introduction

Calorimetry is the experimental technique that allows the determination of the heat transferred between two systems with different temperatures. Basically, it can be applied to obtain the specific heats of the substances, as well as the heats of phase transitions and formation/decomposition processes. The accuracy of the assay of the abovementioned parameters depends on the accuracy of the mass and temperature measurements and the purity of the investigated samples. Sometimes it is necessary to estimate the specific heats for hypothetic materials or materials that are expensive, difficult to synthesize or dangerous for humans and environment. For such cases the heat capacity can be estimated by thermodynamic calculations using the enthalpies and entropies from the thermodynamic databases for pure materials, as in the CALPHAD method. An alternative way, especially when the databases do not contain usable information, is to use the computing methods that can predict the total energy and the vibrational frequencies of a given system. The density functional theory (DFT) is an electronic structure method that considers the electron correlation at a low computational cost and provides accurate results. The high-quality calculations of the vibration frequencies for periodic and nonperiodic systems in various approximations (harmonic, quasiharmonic, and anharmonic) support the calculation of the thermodynamic properties (enthalpy, entropy, and Gibbs free energy) for different composition, pressure P, and temperature T conditions, which can be used to calculate the heat capacity.

In this chapter, we present the theoretical basis of the thermodynamics approach of the heat capacity for crystals for given thermodynamic conditions P and T based on the DFT calculations, having as example six polymorphs of the magnesium hydrides, which are reported in Ref. [1] where the assessments of the thermodynamic properties and pressure–temperature phase diagram of the magnesium hydride polymorphs are done.

Several polymorphs of magnesium hydride were experimentally identified and validated by several electronic structure calculations [1–3]. The knowledge of the various MgH_2 phase stability has led to an increase in research regarding the pressure–temperature phase diagram for the magnesium hydride. Under ambient conditions, the magnesium hydride crystallizes as α-MgH_2 phase, with a rutile-type structure (space group *P42/mnm*) [4]. Bastide et al. found that under high pressure and temperature conditions, the α-MgH2 structure is transformed into β-MgH_2 (space group *Pa-3*) and γ-MgH_2 (space group *Pbcn*); by decreasing the pressure, the β-MgH_2 is transformed into γ-MgH_2 [5].

The magnesium hydride P–T phase diagram based on the thermodynamic calculations [1] shows that in the pressure range 0–1.5 GPa, the α-MgH_2 phase is the most stable and that the γ-MgH_2 is stable below 6.2 GPa. Above this value, the ε-MgH_2 becomes the most stable up to 10 GPa, the maximum pressure considered in the study. However, in the region of 5.5–7.5 GPa, with exception of the cubic phase *c*-MgH_2, all the investigated magnesium hydrides might coexist, since their enthalpies have similar values. The hypothetic *c*-MgH_2 polymorph is the less stable phase, excepting the interval of 0.0–0.4 GPa, where it has enthalpy values comparable with the ε-MgH_2 phase. The cubic polymorph was identified in an experiment as metastable nanocrystals, which transform to γ- and α-MgH_2 [6]. In a subsequent study [7], we predicted that the formation/decomposition curve over all polymorphs is starting from 591.1 K for the low-pressure P = 0.03 GPa, growing with the applied pressure.

The theoretical framework for the thermodynamic calculations presented in the present chapter of the book can be extended to nonperiodic systems (defects, surfaces, interfaces, alloys, amorphous, fluids, and isolated molecules) still using the periodic formalism, but modeling the system in the supercell method, or finite systems (molecules and macromolecules, clusters), where the molecular orbital formalism (specific to the quantum chemistry, which is proper to the isolated systems) is used instead of the crystal orbital formalism (specific to the quantum solid state). For the finite systems, the contributions of the rotation and translation freedom degrees have to be added to the partition function and the free energy of the system.

2. Total energy computation methods

The computer-assisted simulations of the particle systems require: (i) a model for the system, which specifies the chemical species, the position in space (given in Cartesian, internal, or redundant coordinates) and, in case of dynamic treatments, the velocity of each particle; (ii) the method that describes the interactions between the particles and different parameters regarding the calculation method; (iii) different parameters that describe the simulation method (threshold parameters, calculation schemes, and eventually the parallelization technique); and (iv) the parameters and properties that have to be reported by the simulation software.

An ideal crystal can be represented as an indefinitely extended lattice that can be obtained by the translation of a repeated parallelepipedic box, called unit cell. The unit cell is populated with a set of atoms (called atomic basis) that may be arranged in some special points characterized by a set of symmetry operations that is specific for each polymorph of a crystalline substance. The unit cell is characterized by the lengths and mutual orientations of three lattice vectors that delimit the unit cell shape. The solid-state physics is the science that characterizes and classifies the crystals and tries to establish a relation between the nature of the atoms, the structures formed by them, and the crystal properties. The infinite crystal is reduced to the study of the properties of the unit cell, the smallest piece of crystal that preserves the properties of the entire system. The crystals have some local or extended defects, but without losing their global ordering. The crystalline materials can be built based on the experimental structural data that are collected in several online databases or trying to predict it by molecular mechanics and ab-initio calculations. The symmetry of the periodic systems (1D for polymers, 2D for surfaces or films, and 3D for solids) can be used to reduce the computation effort [8].

The equations of state of a given system are obtain by successive approximations having as starting point the time-dependent or the—independent Schrödinger equation associated to electrons and nuclei that form the system. The relativistic contributions to total energy are important for heavy chemical elements and must be considered at least as a perturbation. Due to the relative light mass of the electron compared with the mass of the nuclei, movement of the nuclei and electrons is separated in the frame of the Born-Oppenheimer approximation. The approximation is suitable when the electrons wave function gradient depending of nuclei positions has very small values. The Born-Oppenheimer approximation is not valid when energy values of different electronic states are very close in energy at some nuclear configurations.

For the finite systems (atoms, molecules, clusters) the mono-electronic wave function $\varphi_i\left(\vec{r}\right) = \sum_{\mu=1}^{\varpi} c_{\mu i}\chi_\mu\left(\vec{r}\right)$ is developed in analytic or numeric basis sets $\chi_\mu\left(\vec{r}\right)$, and thus the complex equations are transformed in some treatable ones. In the Linear Combination of Atomic Orbitals (LCAO) approach [9], the functions $\chi_\mu\left(\vec{r}\right)$ are atomic orbitals, and the amplitude of the coefficients $c_{\mu i}$ can be used to interpret the interaction in the system. The radial part of the AO might have different mathematical representations (Slater-type, Gaussian-type, or numeric atom-centered orbitals). In the LCAO approach, the parameters can be decomposed into atomic orbital contributions that can be used to interpret the interaction in the system. The wave functions are called Molecular Orbitals and the corresponding theory, Quantum Chemistry.

For a periodic system (crystal, slabs, surfaces, wires, and tubes), the Born-von Karman boundary condition introduces the expansion of physical quantities to Fourier series. To simplify the solving of the mono-electron Schrödinger equation, the Bloch theorem exploits the translation symmetry of the system and implicitly of the potential, by factorizing the mono-electron wave function as $\varphi_i\left(\vec{r}\right) \to \varphi_{n\vec{k}}\left(\vec{r}\right) = e^{i\vec{k}\vec{r}} u_{n\vec{k}}\left(\vec{r}\right)$, where $\vec{k}$ is a vector defined in the reciprocal space, and n is the band index. Thus, the Bloch theorem indicates the way to reduce the computing of an infinite number of electronic wave functions to a finite number of electron wave functions, as well as the indexation of the electron wave functions by the band index n. The wave functions are called Crystal Orbitals (CO) and the corresponding theory, Quantum Solid-State.

By solving the reduced mono-electronic equation for each $\vec{k}$, a set of energies $\in_{n\vec{k}}$ is obtained. By using the periodicity of the reciprocal space, a polyhedron called Brillouin zone (BZ) may be defined in the reciprocal space. The wave vectors outside the Brillouin zone simply correspond to states that are physically identical to those states within the BZ. For each band n, the energy levels $\in_{n\vec{k}}$ evolve smoothly with the changes in $\vec{k}$, forming a continuum band of states. The electronic wave functions for closed $\vec{k}$-points are very similar, and the integration in the reciprocal space can be reduced to a summation over a grid of k-points [10]. The integrals over k-space converge exponentially with the number of sampling k-points and several recipes are available to compute the sets of spatial $\vec{k}$-points for different symmetries in order to accelerate the convergence of BZ integrations were developed [11]. Due to the partial filling of the energy bands in the case of the metals, the BZ is discontinuous. In this case, the calculations of the integrals over BZ with a denser$\vec{k}$-grid and the broadening of the electronic levels may reduce the magnitude of the errors [12].

The development of the electronic functions using a plane wave (PW) basis set $u_{n\vec{k}}\left(\vec{r}\right) = \sum_{\vec{G}=1}^{\infty} c_{\vec{G}i}\left(n, \vec{k}\right) e^{i\vec{G}\vec{r}}$ is a natural choice for the crystals, as the equations obtained are very similar with those of Nearly Free Electron model [13]. Therefore, the mono-electronic wave function may be written as $\varphi_{i\vec{k}}\left(\vec{r}\right) = e^{i\vec{k}\vec{r}} u_{n\vec{k}}\left(\vec{r}\right) = \sum_{\vec{G}=1}^{\infty} c_{\vec{G}i}\left(n, \vec{k}\right) e^{i\left(\vec{k}+\vec{G}\right)\vec{r}}$. The different terms of the total energy are written as Fourier transforms, thus simplifying the numerical treatments. For the periodic systems, the use of a plane wave basis set in the description of the COs offers a number of advantages, including the simplicity of the basis functions, the absence of basis set superposition error, and the ability to calculate efficiently the forces acting on the atoms. In case of the ionic crystals, the PW method is not efficient because of the high number of plane waves that are required for an accurate description of the wave functions near the ionic core.

Only the electrons that occupy the high energy levels (called valence electrons) are responsible for formation and breaking of the chemical bonds and for the interaction with the low-energy radiation. The rest of the electrons (called core electrons) generally are not affected by the chemical environment and are not as significant as the valence electrons. Thus, treating explicitly only the valence electrons, a further decrease of the computational effort can be achived. The core electrons are emulated by effective core potentials (ECPs) in quantum chemistry LCAO methods or by pseudopotentials (PPs) in the solid-state PW methods [14]. Thus, only the valence electrons are considered in the electronic equations [15]. The core potentials are developed so as to reproduce the energies, as well as the wave function amplitude (outside of a given cutoff radius) of the atomic core wave functions. The PPs that satisfy the condition of the normalization outside of the cutoff radius are called norm-conserving pseudopotentials (NCPP) [16], and those for which this condition is relaxed are called ultrasoft pseudopotentials (USPPs) [17]. The PPs allow one to perform the calculations at a lower energy cutoff. The USPPs are less computationally expensive in comparison with NCPP. For the heavy elements, the relativistic effects can be incorporated in the ECP/PP [18].

2.1 The simulation methods for the electronic ground states

From a broad palette of electronic structure methods, the quantum methods based on the Hartree-Fock Theory (HFT) [19, 20] and the Density Functional Theory (DFT) [20] are the most popular and mostly used. The neglect of the electron correlation in HFT affects the quality of the results, especially the energy-derived properties. The influence of the electron correlation on the band structure and on the cohesion energy in the oxide materials has been the object of a huge number of articles. The electron correlation effects are important for the ionic crystals, as the electron density is localized in a reduced domain around the ionic core. The existing post-Hartee-Fock correction schemes based on the Configuration Interaction, Coupled Cluster Theory, or Møller-Plesset Perturbation Theory are very accurate [21], but too expensive from a computational point of view to be applied to large systems. Some technical difficulties make them very rare in the solid-state software. Fortunately, the methods based on the DFT are good alternatives. In DFT, the total energy is expressed in terms of total electron density, rather than the many-electron wave function specific to HFT. The choice of the exchange-correlation potentials is a matter of trial and error in the DFT, as the method itself does not provide an explicit dependency of the exchange-correlation potentials on the electron density. Developed initially based on the model of the uniform electron distribution in the Local Density Approximation (LDA), the accuracy of the DFT was increased after including additional corrections that consider the variation of the density in the Generalized Gradient Approximation (GGA). Despite the numerous improvements of the DFT methods on the top of GGA (Self-Interaction and Hubbard U corrections, meta-GGA, and hybrid functionals) [22] regarding the overestimation of the band gap for the semiconductors, there are still efforts to find transferable correlation-exchange potentials. The various proposed exchange-correlation potentials do not describe well the weak van der Waals interactions between the two chemical systems. The simplest solution is to treat the dispersion interaction adding an analytic empirical term of London type to the total energy [23].

Due to the similarity of the Hartree-Fock and Kohn-Sham equations, several electronic structure codes work within the framework of both HFT and DFT and can treat the exchange interaction in a hybrid scheme, incorporating the full or partial Fock exchange in the DFT calculations. Both methods can separately treat the two spin orientations up and down of the electrons, in so-called Unrestricted HF in HFT and spin-polarized or spin-density calculations in DFT. This double framework approach allows the study of the electron correlation effects in 0d-3d periodical systems in a unified way.

The several approaches can be applied in order to reduce the calculation effort by approximation of the integrals by simpler formulas or just parameterization as in the Extended Hückel and various Zero Differential Overlap methods [24] in the frame of the HFT or Density Functional Tight Binding (DFTB) in DFT. These methods are called semiempirical methods, as they contain some empirical determined parameters. Generally, the semiempirical methods consider only the valence electrons, the effects of the core electrons being included in the parameterization of the method. The computational effort required by the semiempirical methods is significantly reduced comparing with the ab-initio methods due to the drastically simplification of numeric calculations and of the great reduction of the considered number of electrons. The prices that must be paid are the reduced accuracy and the reduced transferability of the parameters to other chemical systems that those used for parameterization. However, the semiempirical methods can be used as tool for the pre-selection of the materials that can be investigated in the high-throughput computational screening techniques.

In order to further reduce the calculation efforts, the interaction between atoms can be represented by analytical formula developing so-called the empirical force fields (EFFs), where usually the electrons are not explicitly considered. Thus, the electronic freedom degrees are eliminated, and the computing effort is drastically reduced to square of the total number of particles. Further reduction of the number of the freedom degrees can be done by partly or total freezing of the internal geometry of molecules. The parameters of EFF can be obtained by fitting the total energies and forces calculated by ab-initio electronic structure methods [25]. Also, the phonon properties have to be included into the parameterization procedure as reference data for accurate calculations of the thermodynamic properties [26].

2.2 The simulation methods for the electronic excited states

Neither HFT nor DFT is able to treat the electronic excitation and to characterize the electrons in the excited states. The HFT applied to the excited states is equivalent to the electron configuration methods for the ground state. The DFT is constructed based on the ground electronic density and cannot be directly extended to the excited electronic states. The use of multi-reference and the perturbative correlation interaction methods over the Hartree-Fock wave functions is able to characterize the excited states, but they cannot be applied to large systems because of the high computational efforts. Moreover, such methods require very large basis sets, which drastically increase the size of the CPU memory that is necessary to store very large matrices. There are trials to correct for the excited states, the DFT-determined states with the Configuration Interaction [27], Random-Phase Approximation [28], or Machine Learning [29], but such methods are not implemented in the available quantum chemistry or solid-state software.

A more natural approach is to start from the excitation process by the electromagnetic field as an external field. Thus, the HFT and DFT must be reconsidered in the frame of the time-dependent Schrödinger equation. In the case of DFT, an analogous equation to the static Kohn-Sham theorem that states that any expectation value is a functional of the density, and the initial state is established. This formalism is called Time-Dependent Density Functional Theory (TDDFT) [30]. TDDFT has the same problem like the static DFT as the exchange-correlation potential is not defined. The adiabatic approximation treats the exchange-correlation kernel as static, which permits its evaluation from the derivative of the ground state exchange-correlation potential with respect to the density. The simplest choice is the Adiabatic Local Density Approximation, in which the exchange-correlation kernel is calculated from the ground-state LDA functional [31].

The excited electron and the local environment of the electron (hole) behave like a collective excitation called pseudoparticle, which could be treated in the Many-Body Perturbation Theory (MBPT) and the Green's function formalism. The GW formalism uses an similar equation to Kohn-Sham equation that governs the energy and the wave functions of the quasiparticle, only that the exchange-correlation potential is replaced by an integral over the self-energy operator that incorporates all the electron–electron interactions [32]. The mean field is determined from DFT calculations and is used to calculate the GW interaction terms. When in GW only the energy of the quasiparticle is modified, but the wave function is kept unchanged, then the method is simplified (so-called G0W0), and the energy of the quasiparticle is obtained as a first-order HF or KS energy. The corrected energies in the G0W0 method give accurate band gaps for semiconductors or insulators. The introduction of the self-consistency within GW gives better band gaps than G0W0, but with a much higher computational effort. The use of the hybrid functionals for calculation of the initial wave functions is desired, but the GW calculations are very expensive, and thus, the application is limited to the small systems. There is some improvement of the algorithms that allow application of GW method to large systems. The GW approaches can be applied to the neutral excitations, but are not able to consider the charged excitations. The Bethe-Salpeter equation (BSE), which is based on the two-particle Green's functions and the effective two-particle interaction kernel, solves the shortcoming [33]. The kernel can be expressed as sum of the derivatives of the Hartree potential function on the self-energy calculation in GW method. The GW and BSE methods predict more accurately the excitation energies and absorption spectra, compared with DFT methods [34]. Unfortunately, the computational effort increases drastically in the order DFT $<$ GW $<$ BSE. For large systems (more than few hundreds of atoms), even the DFT calculations are prohibited. The semiempirical methods with an accurate parameterization [35] can be applied at reduced computational efforts for large systems in the ground [36] or excited states [37].

2.3 Static and dynamic properties

Total energy is dependent on the relative arrangement of the atoms. Changing continuously the position of the atoms, the energy is continuously modified, and this function energy – coordinate is called potential energy surface (PES). The configuration of atoms that are characterized by minimum or maximum values of the system energy corresponds to an equilibrium or transition states, respectively. Such atomic configurations can be determined by using some mathematical methods that modify the position of the atoms in order to minimize or maximize the total energy, preferentially using the first and second derivatives of the total energy. The derivatives are obtained analytically or numerically. The energy and its derivatives calculated by the electronic structure methods can be used as reference data in the empirical force fields parameterization, in order to reproduce the static and dynamics properties of the atomic systems [38]. The static calculations are very useful to characterize the stability, elastic and electronic properties, the vibration spectra and the thermodynamic properties, and the way of transition from a structure to another.

Ehrenfest theorem establishes the theoretical basis for time evolution of a quantum system. Due to nuclei large masses, the Ehrenfest theorem can be reduce to Newton equation, which rules the atoms movement on the PES. The theory that describes the system time evolution is called Molecular Dynamics (MD). The dynamic properties of the investigated system can be accessed through the MD simulations, which consist of the integration of the Newton equation of each atom of the system. The forces that act on the atoms can be evaluated from electronic structure calculation or can be evaluated by much cheaper empirical force fields. Besides total energy minimization, MD simulations allows the simulation: (i) of the behavior of atoms that form the system (ii) of any kind of chemical species using empirical and quantum force field (iii) of the different statistical ensembles. After saving the trajectories in files and data process, we can: (i) visualize the dynamic of the system; (ii) plot and analyze temperature, pressure, distances, volume, tensile forces, and cell parameters; (iii) calculate velocity correlation factor and vibration intensities; (iv) calculate radial distribution functions and structure factors; (v) calculate various thermodynamics parameters in different thermodynamic ensembles [39].

PES structure and system thermodynamics can be characterized by the heuristic methods and data averaging, which describe the system by very large number of attempts. The most common method is Monte Carlo (MC) based on random-walk movement on PES. The sequences of the events are not related with the time evolution of the system as in case of the MD; they are rather dependent on the chosen algorithm. Because the time evolution of the system is not considered, the MC simulation cannot describe the nonequilibrium processes. Time independence can be an advantage for processes that are taking place on a large timescale or in case of PES with a high "roughness."

2.4 Equation of states and pressure

The calculation of accurate equations of state (EOS), thermodynamic properties, and phase diagrams is a very useful tool in a number of fields, including geophysics [40] and materials research [41]. One of the main advantages of the calculation of pressure and temperature-dependent crystal properties is the ability to investigate the extreme thermodynamic conditions, unattainable by experimental means. Indeed, provided that there are no technical difficulties (e.g., pseudopotential transferability issues [42]), pressure effects can be accounted for simply by compression/expansion of the calculated crystal geometry to smaller/larger volumes compared with the equilibrium volume.

In the present study, the DFT calculations are based on plane waves basis sets techniques combined with ultrasoft pseudopotentials and were carried out with the *Quantum Espresso (QE)* package [43]. The *thermo_pw* package [44] was used to obtain the thermodynamic properties within harmonic and quasi-harmonic approximations. The PBE-GGA [45] exchange-correlation potentials were used in the calculations. We have chosen computational settings to ensure that all investigated properties are well converged. The kinetic energy cutoff was set to 55–60 Ry while the charge density cutoff was set to 300 Ry. The integration over the Brillouin zone (BZ) was performed employing a Monkhorst–Pack mesh with a spacing of 0.04 $Å^{-1}$, as a good balance between the calculation accuracy and the computational effort. The total energy of the polymorphs of the MgH_2 has been minimized function with respect to the unit cell parameters and fractional coordinates of the atoms, preserving the symmetry of the crystal.

The equilibrium volume V_0 and the corresponding energy U_0 can be determined by a full optimization of the lattice parameters and atom fractional coordinates. Furthermore, the volume dependency of the unit cell $U(V)$ can be obtained by scaling the lattice constants with the same factor (isotropic procedure) and calculating the corresponding energy $U(V)$. The minimum of the curve corresponds to the equilibrium point (V_0, U_0). The curve can be fitted by the Murnaghan Equation of States (EOS) [46], which relates by a simple polynomial equation the volume $V$and its equilibrium value V_0, the bulk modulus $B = -V\left(\frac{\partial P}{\partial V}\right)_T$ and its derivative $B' = \left(\frac{\partial K}{\partial P}\right)_T$ by

$$U(V) = U_0 + \frac{B_0 V}{B'}\left[\frac{(V_0/V)^{B\prime}}{B'-1} + 1\right] - \frac{V_0 B_0}{B'-1} \qquad (1)$$

More complex EOSs are available in the literature (see the citations in Refs. [47,48]), but their qualities are similar. The dependence of the energy on the unit cell volume obtained for the different magnesium hydride structures is presented in

Figure 1. The obtained values of the parameters U_O, B_0 and B' by fitting the EOS given by Eq. (1) are presented in **Table 1**. The minimum of the curves gives the equilibrium volume V_O and the equilibrium energy U_O, which are identical with those calculated by a full optimization (lattice parameters and fraction coordinates of the atoms) and are in very good agreement with the experimental values (see Table 1 in Ref. [1]).

At low temperatures, the system energy is only dependent on the system volume, and the pressure can be estimated by

$$P(V) = -\frac{dE}{dV} = \frac{B_0}{B'}\left[\left(\frac{V_0}{V}\right)^{B\prime} - 1\right] \tag{2}$$

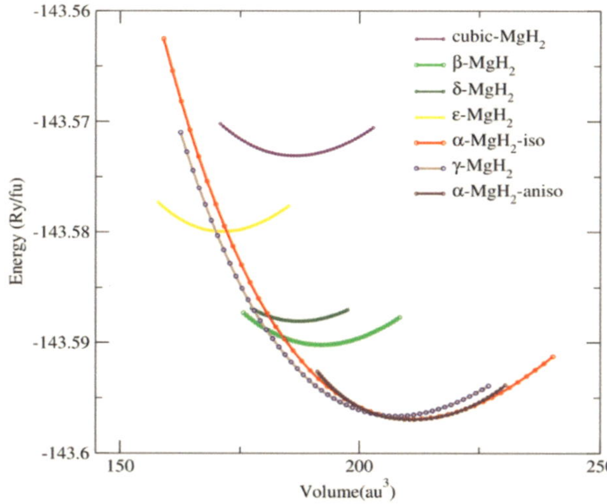

Figure 1.
The energy vs. volume for the polymorphs α, β, γ, δ, ε and cubic of the magnesium hydride obtained by an isotropic procedure. The dependency energy-volume determined by an anisotropic procedure is given for the polymorph α-MgH_2.

Polymorph of MgH_2	Symmetry group	MgH_2 formula unit (f.u.) per unit cell	Total energy U_0 (Ry/f.u.)	Electronic smearing term (Ry/f.u.)	Equilibrium volume V_0 (Å^3)	Bulk modulus B (GPa)	B′	Debye temperature (K)
α (isotropic)	*P42/mnm*	2	−143.5961	0.0014	62.50	51.1	3.59	707.4
α (anisotropic)	*P42/mnm*	2	−143.5961	0.0014	62.48	58.3	4.65	691.2
β	*Pa-3*	4	−143.5902	0.0002	113.67	54.6	2.48	699.9
c	*Fm-3 m*	1	−143.5734	−0.0004	27.55	63.7	1.00	511.2
δ	*Pbca*	8	−143.5880	0.0000	222.23	57.8	1.60	739.9
ε	*Pnma*	4	−143.5801	0.0000	101.48	61.12	4.10	627.5
γ	*Pbcn*	4	−143.5966	0.0001	123.15	47.5	3.48	717.7

Table 1.
The properties of the investigated polymorphs of the MgH_2 obtained from the Murnaghan equation of state. For the polymorph α-MgH_2, the values are presented both for the isotropic and anisotropic volume adjustments. The total energy was corrected by cold smearing procedure, included in Quantum espresso code.

A special care has to be paid when the dependency of energy versus volume is determined in the case of the non-cubic systems, where the lattice parameters and fractional coordinates of the atoms must be optimized for each value of the volume, in order to assure that the energy has the minimum value. Such calculations are done on a grid of lattice parameter under the constraint of a given value of the volume, and those lattice parameters that assure minimum energies are identified. In order to check the effects of the lattice relaxation, we determined the $U(V)$ dependency for the polymorph α-MgH_2 by scaling the lattice constants with the same factor (isotropic procedure) and by a lattice-grid calculation (anisotropic procedure). In the isotropic procedure, nine equidistant scaling factors of the volume between 0.80 and 1.20 were considered. In the anisotropic case, a grid 5×5 of a and c/a, centered at the equilibrium values, with steps of 0.05 Å and 0.02, respectively, was considered. The energy was fitted with quartic polynomials as a function of a and c/a and the pairs $(a,c/a)$ for which the energies have a minimum were identified. The results of the Murnaghan fit are given in **Table 1** and **Figure 1**. It can be seen that in case of α-MgH_2, the isotropic and anisotropic $U(V)$ essentially have the same behavior around the equilibrium volume V_0. The same trend also was observed for the other non-cubic polymorphs of the MgH_2.

3. Phonon calculations in lattice dynamics method: Harmonic approximation

The crystal potential energy can be expanded as a Taylor series [49] by small displacements $u_{\alpha,mI}$ from their equilibrium positions of the atoms I located in the unit cell m, along the Cartesian direction $\alpha = x/y/z$, as

$$U = U_o + U_1 + U_2 + U_3 + \dots \tag{3}$$

where:

U_0- is the static energy of the system in the equilibrium geometry,

$U_1 = \sum_{mI\alpha} \Phi^{\alpha}_{mI} u_{\alpha,mI} = 0$- is zero as the system is in the equilibrium geometry.

The other terms are the n-body crystal potentials (n = 2, 3, ...). The second- and the third-order potentials are

$$U_2 = \frac{1}{2} \sum_{mI\alpha} \sum_{lJ\beta} \Phi^{\alpha\beta}_{mI,lJ} u_{\alpha,mI} u_{\beta,lJ} \tag{4}$$

$$U_3 = \frac{1}{6} \sum_{mI\alpha} \sum_{lJ\beta} \sum_{nK\gamma} \Phi^{\alpha\beta\gamma}_{mI,lJ} u_{\alpha,mI} u_{\beta,lJ} u_{\gamma,nK} \tag{5}$$

where $\Phi^{\alpha\beta}_{mI,lJ}$ and $\Phi^{\alpha\beta\gamma}_{mI,lJ}$ are the harmonic and cubic anharmonic force constants, respectively.

Limiting the expansion to second term, we have the *harmonic approximation*, which is the fundament of the vibrational frequencies calculations. The reduced Hamiltonian$H_{\text{HA}} = \frac{1}{2} \sum_{mI\alpha} m_I \dot{u}^2_{\alpha,mI} + U_2$, which includes the kinetic energy of the atom

I with the mass m_I, is the harmonic Hamiltonian of the system. The problem of N_a atoms per unit cell that moves in a periodic potential is separable and can be solved exactly by diagonalization of the equation of the eigenvalues and eigenvectors

$$\sum_{\beta,J} D_{IJ}^{\alpha\beta}\left(\vec{q}\right) e_{\beta,J}\left(\vec{q},j\right) = \omega^2_{\vec{q},j} e_{\alpha,I}\left(\vec{q},j\right) \tag{6}$$

of the dynamical matrix

$$D_{IJ}^{\alpha\beta}\left(\vec{q}\right) = \frac{1}{\left(m_I m_J\right)^{1/2}} \sum_l \Phi_{0I,lJ}^{\alpha\beta} e^{i\vec{q}\left(\vec{r}_{lJ}-\vec{r}_{0I}\right)} \tag{7}$$

The solutions $\omega_{\vec{q}j}$ and $e_{\alpha,I}\left(\vec{q},j\right)$ are the frequency and polarization vector that correspond to the phonon normal mode of band index j and wave vector $\vec{q}$ in the Brillouin zone. The matrix $D_{IJ}^{\alpha\beta}\left(\vec{q}\right)$ is a hermitic one and its eigenvalues have to be positive. However, sometimes the normal modes might have negative eigenvalues, which correspond to imaginary frequencies. In such a case, the energy decreases along the eigenvector $e_{\alpha,I}\left(\vec{q},j\right)$ and the system becomes unstable.

The phonon frequencies for a unit cell in equilibrium (i.e., the energy is minimized and there are no forces on atoms and no stress in the unit cell) can be calculated from the derivative of the energy (phonon freeze method) [50] or considering the forces of each atom in the frame of the linear response method (Density Functional Perturbation Theory—DFPT) [51]. The phonon occupation number at the equilibrium is given by the Bose-Einstein distribution $n_{\vec{q}j} = \left(\exp\left(\hbar\omega_{\vec{q}j}/k_B T\right) - 1\right)^{-1}$. At absolute zero temperature, the phonon population is zero, at low temperatures, there is a small probability for the phonons to exist, and at high temperatures, the number of phonons increases with temperature. The maximum number of phonon modes is given by the maximum number of freedom degrees $3 \cdot N_a$, where N_a is the total number of atoms in the unit cell of the crystal.

The relation between $\omega_{\vec{q}j}$ and $\vec{q}$ for each mode j, namely $\omega_j = \omega_j\left(\vec{q}\right)$, is called phonon dispersion. The number of the phonon modes with the frequency between ω and $\omega + \Delta\omega$ gives the phonon density of states $g(\omega) = 1/N\sum_{\vec{q}j}\delta\left(\omega - \omega_{\vec{q}j}\right)$, where N is the number of unit cells in the crystal. The normalization factor $1/(3 \cdot N_a)$ is introduced in order to reduce to 1 the integral of $g(\omega)$ over frequency $\int g(\omega)d\omega = 1$.

The DFPT method [51], implemented in the module *thermo_pw* code [44], was used to calculate the phonon spectrum for the MgH_2 polymorphs for each of the volume values on a $6 \times 6 \times 6$ grid of $\vec{q}$-points and Fourier interpolated in the Brilloun Zone. The phonon density of states (DOSs) are computed by integrating the phonon dispersion in the $\vec{q}$-space. We present in **Figure 2** the phonon band structure and DOS for the polymorph α-MgH_2. The phonon spectra are similar with the one obtained by inelastic neutron scattering [52] and by shell-model EFF lattice dynamics calculations [53]. The phonon spectra for the cubic polymorph *c*-MgH_2 show a large number of imaginary frequencies for all nine different volumes. Therefore, this polymorph is not considered for the heat capacity calculations.

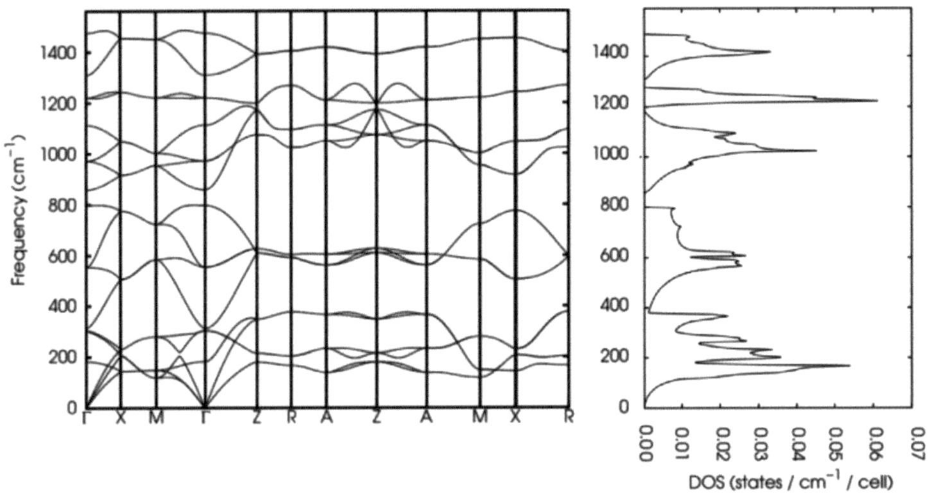

Figure 2.
The phonon dispersion (left) and density of states (right) computed for the polymorph α-MgH_2.

4. The canonical partition function and the Helmholtz free energy

The canonical partition function of the system is the defined as $Z(N,V,T) = \sum_i e^{-\frac{E_i(N,V)}{k_B T}}$, where the summation is done over all the states that are characterized by the energies E_i of the system formed by N particles that occupy the volume V and is kept under the temperature T. k_B is the Boltzmann's constant.

The thermodynamic parameters of the system can be defined as

$$U = k_B T^2 \left(\frac{\partial \ln Z}{\partial T}\right)_{N,V} \quad \text{– average potential energy} \tag{8}$$

$$S = k_B \ln Z + U/T \text{ – entropy} \tag{9}$$

$$H = U + P \cdot V \text{ – enthalpy} \tag{10}$$

$$F = U - T \cdot S \text{ – Helmholtz free energy} \tag{11}$$

In the relation above k_B is the Boltzmann constant, T is the temperature, P is the pressure, and V is the volume of the unit cell. Enthalpy, defined as $H = U + P \cdot V$, is a measure of a system capacity to release heat as a nonmechanical work. The most fundamental thermodynamic parameter is the Gibbs free energy, which is defined as $G(P,T) = F + p \cdot V = H - T \cdot S$.

Based on the Helmholtz free energy, other important thermodynamic parameters can be defined:

- the bulk modulus $B_T = V\left(\frac{\partial^2 F}{\partial V^2}\right)_T$, which describes the resistance of a material to the compression,

- the isochoric heat capacity $C_V = -T\left(\frac{\partial^2 F}{\partial T^2}\right)_V$, which is the amount of heat per unit mass of material that is required to raise the temperature by one unit under the condition of the constant volume,

- the isobaric heat capacity $C_P = \left(\frac{\partial H}{\partial T}\right)_P = C_V - T\left(\frac{\partial V}{\partial T}\right)_P^2\left(\frac{\partial P}{\partial V}\right)_T$, and the volumetric thermal expansion as $\alpha_V = \frac{1}{V}\left(\frac{\partial V}{\partial T}\right)_P$. Generally, the energy of the system can be decomposed based on the different degrees of freedom as the electronic and nuclear (translation, rotation, and vibration) motions, domains contributions (bulk, surface, interface, and defects), or phenomena (magnetism, irradiation). Some hybrid contributions have to be added in case of not completely separation of the different freedom degrees (as example, the electron–phonon coupling). In the case of the crystalline systems, the translation and rotation motions of the atomic basis are not present, and their contributions are not included in the structure of the Helmholtz free energy. Furthermore, the product PV is very small in comparison with the other terms, and usually it is neglected. Therefore, usually the stability of the crystals is characterized by the Helmholtz free energy, rather than the Gibbs energy.

For a nonmagnetic ideal crystal, the Helmholtz free energy can be written as:

$$F(V,T) = U_0(V) + F_{el}(V,T) + F_{phon}(V,T) \tag{12}$$

where the first term in Eq. (12) corresponds to the total energy, the second term is the contribution of the electronic excitation, and the last term is given by the nuclear vibrational motion. The first and third terms can be calculated both by the electronic structure or empirical force fields methods and the second one just by electronic structure methods.

The electronic structure methods typically do not take explicitly into account the temperature effects on the atomic ground state and the computed properties correspond to T = 0 K. However, for the incomplete occupied bands, like in case of the metals, the occupation probability of the electronic levels $\in$ is described by the Fermi-Dirac distribution $f(\in,T) = \left(\exp\left(\frac{\in - E_F}{k_B T}\right) + 1\right)^{-1}$, where the E_F is the Fermi level, the last occupied electronic level, with the value given by the normalization condition $\int f(\in,T)d\in = 1$. The energy due to the electronic excitation is $U_{el}(V,T) = \int n(V,\in)f\in d\in - \int n(V,\in)\in d\in$, where $n(V,\in)$ is the electronic density of states computed for the volume V. The corresponding electronic entropy is $S_{el}(V) = -gk_B \int[f(V,T)\ln f(V,T) + (1-f(V,T))\ln(1-f(V,T))]$.

where g is equal to 1 for collinear spin polarized or 2 for non-spin-polarized systems. The electronic Helmholtz free energy is $F_{el}(V,T) = U_{el}(V,T) - TS_{el}(V,T)$. The electronic excitation term is calculated based on the Mermin theorem [54]. $F_{el}$is important for the metallic systems, especially at high temperatures. Usually it is neglected in the other cases. However, the partial occupation of the electronic level can be used also for the nonmetallic systems in the smearing scheme [55] that is useful to accelerate the self-consistent field convergence, mixing the occupied and unoccupied states that are clearly separated at T = 0 K. The contribution of the smearing to the electronic energy is refereed as Mermin free energy. The smearing scheme was applied to perform the calculations for the investigated polymorphs of the magnesium hydride. It is efficient in the accelerating the SCF convergence, inducing a very low correction to the total energies (see **Table 1**). The electronic contribution to the electronic Helmholtz energy is negligible except for the cubic polymorph c-MgH_2, which has metallic properties.

The third term in Eq. (12) is the contribution of the phonon vibration and is dependent on temperature. The partition function corresponding to the phonon is

$Z_{vib} = \prod_{\vec{q},j}\left(\sum_{n=0}^{\infty} e^{-\left(n+\frac{1}{2}\right)\frac{\hbar\omega_{\vec{q}j}}{k_BT}}\right)$ and the vibrational Helmholtz energy in the *harmonic approximation* is depending on temperature only by the phonon contribution calculated for the equilibrium volume V_{eq}

$$F_{HA}\left(V_{eq},T\right) = U_0(V) + \frac{1}{2}\sum_{\vec{q},j}\hbar\omega_{\vec{q},j} + k_BT\sum_{\vec{q},j}\log\left(1 - e^{-\frac{\hbar\omega_{\vec{q},j}}{k_BT}}\right) \quad (13)$$

For low temperatures, the amplitude of the vibrations is reduced and the harmonic approximation is valid for almost all the cases. For N_a atoms per unit cell, the isochoric heat capacity is

$$C_V = -T\left(\frac{\partial^2 F}{\partial T^2}\right)_V = k_B N_a \sum g(\omega)\left(\frac{\hbar\omega}{k_BT}\right)^2 \frac{\exp\left(\hbar\omega_{\vec{q},j}/k_BT\right)}{\exp\left(\hbar\omega_{\vec{q},j}/k_BT\right) - 1} \quad (14)$$

5. Computational thermodynamics in quasi-harmonic approximation

The neglect of anharmonicity by the truncation of the third term in the development of the total energy leads to well-known unphysical behavior [56]: the zero thermal expansion, infinite thermal conductivity, and phonon lifetime. The inclusion of temperature effects, primarily related to the vibrational degrees of freedom inside the crystal, is more delicate. There are essentially two mainstream ways of incorporating temperature in a theoretical calculation: Molecular Dynamics [57] and Monte Carlo [58] simulations, and the *quasiharmonic approximation* (*QHA*) [59]. The former techniques are ideally suited for situations close to the classical limit, at temperatures close to or including the melting temperature. The latter is assuming the harmonic approximation at any given crystal geometry, even if it does not correspond to the equilibrium structure. Plenty of examples of the success of *QHA* in the prediction of thermodynamic properties and phase stability of solids can be found in the literature [60].

In QHA, the phonon calculations are done for several volumes $\omega_{\vec{q},j}(V)$ and the Helmholtz free energy becomes

$$F_{QHA}(V,T) = U_0(V) + \frac{1}{2}\sum_{\vec{q},j}\hbar\omega_{\vec{q},j}(V) + k_BT\sum_{\vec{q},j}\log\left[1 - \exp\left(-\frac{\hbar\omega_{\vec{q},j}(V)}{k_BT}\right)\right] \quad (15)$$

or by the grouping of the terms [61].

$$F_{QHA}(V,T) = U_{cold}(V) + F_{th}(V) \quad (16)$$

where $U_{cold}(V) = U_0(V) + \frac{1}{2}\sum_{\vec{q},j}\hbar\omega_{\vec{q},j}(V)$ is the cold potential energy (T = 0 K), and $F_{th}(V) = k_BT\sum_{\vec{q},j}\log\left(1 - \exp\left(-\frac{\hbar\omega_{\vec{q},j}(V)}{k_BT}\right)\right)$ is the thermal component of the Helmholtz free energy given by the phonons.

The accurate calculation of the phonon frequency requires the energy evaluation by electronic structure methods, but also the usage of some empirical force fields might give good results [62]. At very low temperature, the thermal part of the phonon contribution is negligible and the entropy does not contribute to the Gibbs energy as the product TS is also negligible. In these conditions, the Gibbs energy of the crystal becomes $G \cong U_{cold} + PV$. The product PV is very small comparing with the total energy and the Gibbs energy is approximated with the cold energy. The anharmonic effects become significant for the magnesium hydride just above 800 K [63]. Therefore, we do not consider here the anharmonic effects on the thermodynamic properties.

Alternatively to the static calculations of the phonons in the harmonic approximation, the anharmonic phonon spectra can be calculated by Molecular Dynamics simulations [64, 65]. The phonon dispersion of a given phonon wave vector $\vec{q}$ can be evaluated from MD simulations by computing the Fourier transforms of velocity–velocity correlation functions $g_{\vec{q}}(\omega, T) = \int e^{i\omega t} \sum_{I=1}^{N_a} e^{i\vec{q}\vec{R}_I} \frac{\langle \vec{v}_I(t)\vec{v}_I(0)\rangle_T}{\langle \vec{v}_I(0)\vec{v}_I(0)\rangle_T} dt$, where $\vec{v}_I(t)$ is velocity of each atom I at time t, and $\vec{R}_I$ is the lattice position of the same atom. The angular brackets represent the time average for a given temperature T considered during the MD simulations. The phonon dispersions are explicitly temperature-dependent and include also the anharmonic contributions. The MD calculations fully consider the anharmonic effects, but do not consider the ZPE effects, as they follow the classical statistical mechanics [66]. Therefore, a quantum correction has to be considered, especially for low temperatures [67].

6. Computational thermodynamics

6.1 Thermodynamic modeling

Considering the temperature effects on the volume the pressure can be formulated as

$$P(V, T) = -\left(\frac{\partial F}{\partial V}\right)_T = P_{stat}(V, 0\text{ K}) + P_{phon}(V, T) + P_{ae}(V, T) \tag{17}$$

where $P_0(V, 0\text{ K})$ is the static pressure computed by Eq. (2), $P_{phon}(V, T)$ is phonon contribution to the pressure, and $P_{ae}(V, T)$ is the anharmonic and electronic thermal pressure. The thermal pressure is calculated as the derivative of the thermal free energy

$$P_{phon} = -\left(\frac{\partial F_{phon}}{\partial V}\right)_T = -\sum_{\vec{q}j}\left(\frac{1}{2} + \frac{\hbar\omega_{\vec{q}j}}{e^{\frac{\hbar\omega_{\vec{q}j}}{k_B T}} - 1}\right)\hbar\frac{d\omega_{\vec{q}j}}{dV} = -\sum_{\vec{q}j}\gamma_{\vec{q}j}\left(\frac{1}{2} + \frac{\hbar\omega_{\vec{q}j}}{e^{\frac{\hbar\omega_{\vec{q}j}}{k_B T}} - 1}\right) \tag{18}$$

where $\gamma_{\vec{q}j} = -\frac{V}{\omega_{\vec{q}j}}\frac{\partial \ln \omega_{\vec{q}j}}{\partial V} = -\frac{V}{2\omega_{\vec{q}j}^2}\left\langle e_{\vec{q}j}\left|\frac{\partial D(\vec{q})}{\partial V}\right|e_{\vec{q}j}\right\rangle$ is the so-called mode Grüneisen parameter, which characterizes the volume dependences of the frequency of the mode

$\left(\vec{q},j\right)$. Similarly to Eq. (2), the pressure $P(V) = -\left(\frac{dF}{dV}\right)_T$ can be calculated for each considered volume of the unit cell based on the equations of state of Murnaghan type [46]. The enthalpy can be calculated as $H = U_0(V) + PV$.

6.2 Debye method

The Debye model [68] considers a simple form for the DOS of the vibration modes $g(\omega) = C \cdot \omega^2 \cdot \Theta(\omega - \omega_D)$, where Θ is the Heaviside step function, and $C = 9N_a/\omega_D^3$ is a constant determined from the condition $C\int g(\omega)d\omega = 3N_a$. Thus, the phonon modes are populated just below ω_D, which is called Debye frequency. The phonon Helmholtz free energy and the isochoric heat capacity are

$$F_{phon}^{Debye}(T) = N_a k_B T\left(\frac{9T_D}{8T} + 3\ln\left(1 - e^{-\frac{T_D}{T}}\right) - D\left(\frac{T_D}{T}\right)\right) \tag{19}$$

where the function $D\left(\frac{T_D}{T}\right) = \frac{3}{(T_D/T)^3}\int_0^{T_D/T} \frac{x^3}{e^x - 1}dx$ is the Debye integral and the parameter $T_D = \frac{\hbar}{k_B}\left[6\pi^2 V^{1/2} N_a\right]^{1/3} f(\sigma)\sqrt{\frac{B_s}{M}}$ is the Debye Temperature, which can be interpreted as the temperature at which each mode below the highest-frequency mode ω_D is excited. The Debye model allows the calculation of the thermodynamic parameters avoiding the phonon calculations if the Debye temperature can be determined from experimental or theoretical elastic constants [69] or from the value of the melting temperature [70]. The estimated Debye temperatures for T = 300 K are given in the **Table 1**, for all the polymorphs.

6.3 Heat capacity: Grüneisen parameter

The weighted average heat capacity of the individual phonon modes

$$\gamma = \frac{\sum_{\vec{q},j} \gamma_{\vec{q},j} C_V\left(\vec{q},j\right)}{\sum_{\vec{q},j} C_V\left(\vec{q},j\right)} \tag{20}$$

is the total Grüneisen parameter, where

$$C_V\left(\vec{q},j\right) = k_B\left(\frac{\hbar\omega_{\vec{q},j}}{k_B T}\right)^2\left[\exp\left(\frac{\hbar\omega_{\vec{q},j}}{k_B T}\right) - 1\right]^{-2}\exp\left(\frac{\hbar\omega_{\vec{q},j}}{k_B T}\right) \tag{21}$$

is the contribution of each vibration mode $\left(\vec{q},j\right)$ to the isochoric heat capacity

$$C_V = \sum_{\vec{q},j} C_V\left(\vec{q},j\right) \tag{22}$$

The constant volume (isochoric) heat capacity was obtained from the quasi-harmonic phonon frequencies calculated at each fixed volume. The isochoric heat capacity at each temperature is then obtained interpolating at the temperature-dependent volume values obtained at each temperature from the minimization of the free energy.

The constant pressure (isobaric) heat capacity is obtained as

$$C_P = C_V + TV\beta^2 B_T \tag{23}$$

where $B_T(V,T) = \frac{1}{V}\left(\frac{\partial^2 F(V,T)}{\partial V^2}\right)_T$ is the isothermal bulk modulus, which can be calculated as a function of temperature [61].

In **Figure 3**, the predicted isochoric and isobaric heat capacities of the magnesium hydride polymorphs are shown to have similar behavior, with deviations at 800 K below 3.81 and 3.44 J mol^{-1} K^{-1}, respectively. The isochoric and isobaric heat capacities have the same values bellow 400 K. The predicted heat capacities for the polymorph α-MgH_2 are in excellent agreement with the experimental data [52, 71, 72] despite that the calculations are done for ideal crystalline systems, real sample are generally polycrystalline and contain a lot of defects depending on the preparation procedure. The iso- and anisotropic calculated isochoric heat capacities for α-MgH_2 are identical while those for the isobaric heat capacity are almost identical up to 550 K, which is the decomposition temperature of the crystal; even at the higher temperatures, the iso- and aniso- differences are marginal. Similar results regarding the isotropic heat capacities are obtained for the other non-cubic polymorphs. Therefore, we conclude that the isotropic treatment by uniform contraction/expansion of the unit cell of the considered magnesium hydride polymorphs introduces negligible effects on the calculated isochoric heat capacities, below their decomposition temperatures. The electronic contribution to the heat capacity was found to be negligible.

The heat capacity in the Debye model is $C_V^{Debye} = 3N_a k_B\left(4D\left(\frac{T_D}{T}\right) - \frac{3T_D/T}{e^{T_D/T}-1}\right)$. The formula is valid at low temperatures, predicting correctly the temperature dependence of the heat capacity as being proportional to the temperature at power 3, and recovers at high temperatures the Dulong–Petit law, which states that the heat capacity is proportional with the number of atoms per unit cell $3N_a k_B$ at high temperatures.

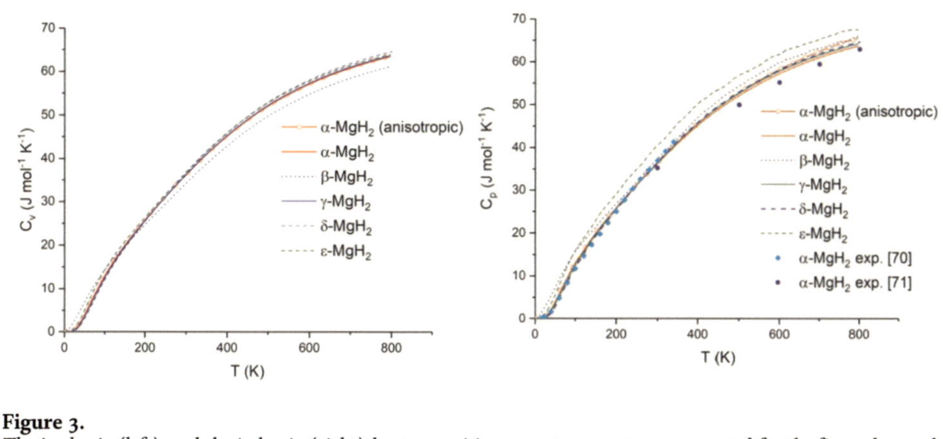

Figure 3.
The isochoric (left) and the isobaric (right) heat capacities versus temperature, computed for the five polymorphs of the magnesium hydride α, β, δ, ε, and γ by the isotropic procedure. In addition, the C_v versus T obtained by the anisotropic procedure is presented for α-MgH_2 (red open circle). The experimental data available for α-MgH_2 are indicated by full symbols.

The Molecular Dynamics and Monte Carlo simulations in tandem with the thermodynamic integration [73] and free energy perturbation [74] methods can be used to calculate the Helmholtz free energy and other thermodynamic properties. The electronic structure methods or well-parameterized EFF can be used in MD simulations to the vibration DOS and to calculate the thermodynamic properties such as heat capacity [75]. The heat capacity can be directly evaluated in MD or MC simulations as $C_{V/P} = \frac{\langle E^2 \rangle - \langle E \rangle^2}{k_B T^2}$, where E is total energy calculated at each simulation step, and the brackets indicate the average over the sequence of configurations produced during the MD or MC steps. Depending on the used thermodynamic ensemble, canonic (N_a, V, T constant) or the isothermal-isobaric (N_a, P, T constant), the isochoric C_v, and isobaric C_p heat capacity, respectively, are calculated. The MD and MC simulations involve the evaluation of the energy and interatomic forces for a very large number of atomic configurations and are not practical to involve electronic structure methods, especially in the case of large systems.

7. Conclusions

The electronic structure and empirical force fields methods are discussed shortly, emphasizing the calculation of the total energy and its derivatives, as well as the phonon spectra. The methods for the calculation of the Helmholtz, Enthalpy, and Gibbs energy are described. The heat capacities of several polymorphs of the magnesium hydride are investigated by density functional theory within the frame of the quasi-harmonic approximation. The predicted temperature variation of the heat capacity for the polymorph of the magnesium hydride α-MgH_2 is in good agreement with the experimental data. We assess the heat capacity for several other polymorphs of MgH_2, for which there is no available experimental or theoretical information.

Acknowledgements

The work was supported by a grant of the Romanian Ministry of Education and Research, CCCDI – UEFISCDI, project number PN-III-P2-2.1-PED-2019-4816, within PNCDI III. CV and AV acknowledge the financial support for mutual visits provided on the basis of the inter-academic exchange agreement between the Bulgarian Academy of Science and the Romanian Academy.

Conflict of interest

The authors declare that the research was conducted in the absence of any commercial or financial relationships that could be construed as a potential conflict of interest.

Author details

Viorel Chihaia[1]*, Valentin Alexiev[2] and Hasan S. AlMatrouk[3]

1 Institute of Physical Chemistry Ilie Murgulescu, Romanian Academy, Bucharest, Romania

2 Institute of Catalysis, Bulgarian Academy of Sciences, Sofia, Bulgaria

3 Kuwait Institute for Scientific Research, Safat, Kuwait City, Kuwait

*Address all correspondence to: vchihaia@icf.ro

References

[1] AlMatrouk HS, Chihaia V, Alexiev V. Density functional study of the thermodynamic properties and phase diagram of the magnesium hydride. Calphad. 2018;**60**:7-15. DOI: 10.1016/j.calphad.2017.11.001

[2] Vajeeston P, Ravindran P, Hauback BC, Fjellvåg H, Kjekshus A, Furuseth S, et al. Structural stability and pressure-induced phase transitions in MgH2. Physical Review B. 2006;**73**:224102-224108. DOI: 10.1103/PhysRevB.73.224102

[3] Yartys VA et al. Magnesium based materials for hydrogen based energy storage: Past, present and future. International Journal of Hydrogen Energy. 2019;**44**:7809-7859. DOI: 10.1016/j.ijhydene.2018.12.212

[4] Hakim K, Rivoldini A, Van Hoolst T, Cottenier S, Jaeken J, Chust T, et al. A new ab initio equation of state of hcp-Fe and its implication on the interior structure and mass-radius relations of rocky super-earths. Icarus. 2018;**313**:61-78. DOI: 10.1016/j.icarus.2018.05.005

[5] Bastide JP, Bonnetot B, Letoffe JM, Claudy P. Polymorphisme de l'hydrure de magnesium sous haute pression. Materials Research Bulletin. 1980;**15**:1215-1224. DOI: 10.1016/0025-5408(80)90024-0 ibid. 1980;15:1779-1787. DOI: 10.1016/0025-5408(80)90197-X

[6] El-Eskandarany MS, Banyan M, Al-Ajmia F. Discovering a new MgH2 metastable phase. RSC Advances. 2018;**8**:32003-32008. DOI: 10.1039/C8RA07068G

[7] AlMatrouk HS, Al-Ajmi F, Do NS, Chihaia V, Alexiev V. The pressure-temperature phase diagram assessment for magnesium hydride formation/decomposition based on DFT and CALPHAD calculations. Modern Approaches on Material Science (MAMS). 2021;**4**:467-478. DOI: 10.32474/MAMS.2021.04.000180

[8] Orlando R, De La Pierre M, Zicovich-Wilson CM, Erba A, Dovesi R. On the full exploitation of symmetry in periodic (as well as molecular) self-consistent-field ab initio calculations. The Journal of Chemical Physics. 2014;**141**:104108-104109. DOI: 10.1063/1.4895113

[9] Astier M, Pottier N, Bourgoin JC. Linear-combination-of-atomic-orbitals, self-consistent-field method for the determination of the electronic structure of deep levels in semiconductors. Physical Review B. 1979;**19**:5265-5276. DOI: 10.1103/PhysRevB.19.5265

[10] Folland NO. Finite-sum approximations to cubic Brillouin-zone integrals. Physical Review B. 1980;**22**:3669-3677. DOI: 10.1103/PhysRevB.22.3669

[11] Monkhorst HJ, Pack JD. Special points for Brillouin-zone integrations. Physical Review B. 1976;**13**:5188-5192. DOI: 10.1103/PhysRevB.13.5188

[12] Methfessel M, Paxton AT. High-precision sampling for Brillouin-zone integration in metals. Physical Review B. 1989;**40**:3616-3621. DOI: 10.1103/PhysRevB.40.3616

[13] Francis GP, Payne MC. Finite basis set corrections to total energy pseudopotential calculations. Journal of Physics: Condensed Matter. 1990;**2**:4395-4404. DOI: 10.1088/0953-8984/2/19/007

[14] Schwerdtfeger P. The pseudopotential approximation in electronic structure theory.

ChemPhysChem. 2011;**12**:3143-3155. DOI: 10.1002/cphc.201100387

[15] Kresse G, Hafner J. Norm-conserving and ultrasoft pseudopotentials for first-row and transition elements. Journal of Physics: Condensed Matter. 1994;**6**:8245-8257. DOI: 10.1088/0953-8984/6/40/015

[16] Bachelet GB, Hamann DR, Schlüter M. Pseudopotentials that work: From H to Pu. Physical Review B. 1982;**26**:4199-4228. DOI: 10.1103/PhysRevB.26.4199. Erratum. ibidem, 1984;29:2309-2309. DOI: 10.1103/PhysRevB.29.2309

[17] Vanderbilt D. Soft self-consistent pseudopotentials in a generalized eigenvalue formalism. Physical Review B, 1990;41:7892-7895(R). DOI: 10.1103/PhysRevB.41.7892

[18] Elsasser C, Takeuchi N, Ho KM, Chan CT, Braun P, Fahnle M. Relativistic effects on ground state properties of 4d and 5d transition metals. Journal of Physics: Condensed Matter. 1990;**2**: 4371-4394. DOI: 10.1088/0953-8984/2/19/006

[19] Pisani C, Dovesi R, Roetti C. In: Berthier G et al., editors. Hartree-Fock Ab Initio Treatment of Crystalline Systems. Part of Lecture Notes in Chemistry. Vol. 48. Berlin Heidelberg: Springer-Verlag; 1988. DOI: 10.1007/978-3-642-93385-1 ISBN-13: 978-3-540-19317-3

[20] Kantorovich L. Quantum Theory of the Solid State: An Introduction. Fundamental Theories of Physics vol. 136. A Van Der Merwe. 2004 Springer Science+Business Media Dordrecht. DOI 10.1007/978-1-4020-2154-1. ISBN 978-1-4020-2153-4

[21] Townsenda J, Kirklanda JK, Vogiatzis KD. Post-Hartree-Fock methods: configuration interaction, many-body perturbation theory, coupled-cluster theory. In: Blinder SM, House JE, editors. Mathematical Physics in Theoretical Chemistry. Amsterdam, Netherlands: Elsevier; 2019. pp. 63-117. ISBN: 978-0-12-813651-5

[22] Csonka GI, Perdew JP, Ruzsinszky A, Philipsen PHT, Lebègue S, Paier J, et al. Assessing the performance of recent density functionals for bulk solids. Physical Review B. 2009;**79**:155107. DOI: 10.1103/PhysRevB.79.155107

[23] Steinmann SN, Corminboeuf C. Comprehensive benchmarking of a density-dependent dispersion correction. Journal of Chemical Theory and Computation. 2011;7:3567-3577. DOI: 10.1021/ct200602x

[24] Parr RG. A Method for estimating electronic repulsion integrals over LCAO MOs in complex unsaturated molecules. The Journal of Chemical Physics. 1952; **20**:1499-1499. DOI: 10.1063/1.1700802

[25] Gale JD, Wright K. Lattice dynamics from force-fields as a technique for mineral physics. Reviews in Mineralogy and Geochemistry. 2010;**71**:391-411. DOI: 10.2138/rmg.2010.71.18

[26] Rohskopf A, Seyf HR, Gordiz K, Tadano T, Henry A. Empirical interatomic potentials optimized for phonon properties. npj Computational Materials. 2017;**3**:27-27. DOI: 10.1038/s41524-017-0026-y

[27] Marian CM, Heil A, Kleinschmidt M. The DFT/MRCI method. WIREs Computational Molecular Science. 2019; **9**(e1394):1-31. DOI: 10.1002/wcms.1394

[28] Ziegler T, Krykunov M, Autschbach J. Derivation of the RPA (random phase approximation) equation

of ATDDFT (adiabatic time dependent density functional ground state response theory) from an excited state variational approach based on the ground state functional. Journal of Chemical Theory and Computation. 2014;**10**:3980-3986. DOI: 10.1021/ct500385a

[29] Westermayr J, Marquetand P. Machine learning for electronically excited states of molecules. Chemical Reviews. 2021;**121**:9873-9926. DOI: 10.1021/acs.chemrev.0c00749

[30] Runge E, Gross EKU. Density-functional theory for time-dependent systems. Physical Review Letters. 1984; **52**:997-1000. DOI: 10.1103/PhysRevLett.52.997

[31] Helbig N, Fuks JI, Casula M, Verstraete MJ, Marques MAL, Tokatly IV, et al. Density functional theory beyond the linear regime: Validating an adiabatic local density approximation. Physical Review A. 2011; **83**:032503-032505. DOI: 10.1103/PhysRevA.83.032503

[32] Hedin L. New method for calculating the one-particle Green's function with application to the electron-gas problem. Physics Review. 1965;**139**:A796-A823. DOI: 10.1103/PhysRev.139.A796

[33] Leng X, Jin F, Wei M, Ma Y. GW method and Bethe–Salpeter equation for calculating electronic excitations. WIREs Computational Molecular Science. 2016; **6**:532-550. DOI: 10.1002/wcms.1265

[34] Reining L. The GW approximation: content, successes and limitations. WIREs Computational Molecular Science. 2018;**8**:e1344-e1326. DOI: 10.1002/wcms.1344

[35] Christensen AS, Kubař T, Cui Q, Elstner M. Semiempirical quantum mechanical methods for noncovalent interactions for chemical and biochemical applications. Chemical Reviews. 2016;**116**:5301-5337. DOI: 10.1021/acs.chemrev.5b00584

[36] Dral PO, Wu X, Spörkel L, Koslowski A, Thiel W. Semiempirical Quantum-chemical orthogonalization-corrected methods: Benchmarks for ground-state properties. Journal of Chemical Theory and Computation. 2016;**12**:1097-1120. DOI: 10.1021/acs.jctc.5b01047

[37] Tuna D, Lu Y, Koslowski A, Thiel W. Semiempirical Quantum-chemical orthogonalization-corrected methods: Benchmarks of electronically excited states. Journal of Chemical Theory and Computation. 2016;**12**:4400-4422. DOI: 10.1021/acs.jctc.6b00403

[38] Sami S, Menger MFSJ, Faraji S, Broer R, Havenith RWA. Q-Force: Quantum Mechanically Augmented Molecular Force Fields. Journal of Chemical Theory and Computation. 2021;**17**:4946-4960. DOI: 10.1021/acs.jctc.1c00195

[39] Dove MT. Introduction to Lattice Dynamics. Cambridge: Cambridge University Press; 1993. pp. 179-194. ISBN: 9780521392938

[40] Wentzcovitch RM, Yu YG, Wu Z. Thermodynamic properties and phase relations in mantle minerals investigated by first principles quasiharmonic theory. In: Theoretical and Computational Methods in Mineral Physics: Geophysical Applications, Mineral. Chantilly, VA: Mineralogical Society of America; 2010. pp. 59-98. DOI: 10.1515/9781501508448 ISBN: 9780939950850

[41] Errandonea D, Ferrer-Roca C, Martinez-García S, Segura A, Gomis A, Muñoz A, et al. High-pressure x-ray diffraction and ab initio study of

Ni2Mo3N, Pd2Mo3N, Pt2Mo3N, Co3Mo3N, and Fe3Mo3N: Two families of ultra-incompressible bimetallic interstitial nitrides. Physical Review B. 2010;**82**:174105. DOI: 10.1103/PhysRevB.82.174105

[42] Porezag D, Pederson MR, Liu AY. Importance of nonlinear core corrections for density-functional based pseudopotential calculations. Physical Review B. 1999;**60**:14132-14139. DOI: 10.1103/PhysRevB.60.14132

[43] Giannozzi P et al. QUANTUM ESPRESSO: A modular and open-source software project for quantum simulations of materials. Journal of Physics: Condensed Matter. 2009;**21**: 395502. DOI: 10.1088/0953-8984/21/39/395502. www.quantum-espresso.org [Accessed on 21 November, 2021]

[44] Available from: https://dalcorso.github.io/thermo_pw/ [Accessed on 21 November, 2021]

[45] Perdew JP, Burke K, Ernzerhof M. Generalized gradient approximation made simple. Physical Review Letters. 1996;**77**:3865-3868. DOI: 10.1103/PhysRevLett.77.3865 Erratum: ibidem 1997;78:1396. DOI: 10.1103/PhysRevLett.78.1396

[46] Murnaghan FD. The compressibility of media under extreme pressures. Proceedings of the National Academy of Sciences of the United States of America. 1944;**30**:244-247. DOI: 10.1073/pnas.30.9.244

[47] Cohen RE, Gülseren O, Hemley RJ. Accuracy of equation-of-state formulations. American Mineralogist. 2000;**85**:338-344. DOI: 10.2138/am-2000-2-312

[48] Otero-de-la-Roza A, Luaña V. Gibbs2: A new version of the quasi-harmonic model code. I. Robust treatment of the static data. Computer Physics Communications. 2011;**182**: 1708-1720. DOI: 10.1016/j.cpc.2011.04.016

[49] Togo A, Tanaka I. First principles phonon calculations in materials science. Scripta Materialia. 2015;**108**:1-5. DOI: 10.1016/j.scriptamat.2015.07.021

[50] Kresse G, Furthmuller J, Hafner J. Ab initio force constant approach to phonon dispersion relations of diamond and graphite. Europhysics Letters. 1995; **32**:729-734. DOI: 10.1209/0295-5075/32/9/005

[51] Baroni S, de Gironcoli S, Corso AD. Phonons and related crystal properties from density-functional perturbation theory. Reviews of Modern Physics. 2001;**73**:515-562. DOI: 10.1103/RevModPhys.73.515

[52] Kolesnikov AI, Antonov VE, Efimchenko VS, Granrotha G, Klyamkin SN, Levchenko AV, et al. Neutron spectroscopy of magnesium dihydride. The Journal of Alloys and Compounds. 2011;**509**:S599-S603. DOI: 10.1016/j.jallcom.2010.10.156

[53] Lasave J, Dominguez F, Koval S, Stachiotti M, Migoni RL. Shell-model description of lattice dynamical properties of MgH2. Journal of Physics: Condensed Matter. 2005;**17**:7133-7141. DOI: 10.1088/0953-8984/17/44/006

[54] Mermin ND. Thermal properties of the inhomogeneous electron gas. Physics Review. 1965;**137**:A1441-A1443. DOI: 10.1103/PhysRev.137.A1441

[55] Hofmann OT, Zojer E, Hörmann L, Jeindl A, Maurer RJ. First-principles calculations of hybrid inorganic–organic interfaces: From state-of-the-art to best practice. Physical Chemistry Chemical

Physics. 2021;**23**:8132-8180. DOI: 10.1039/d0cp06605b

[56] Ashcroft NW, Mermin ND. Solid State Physics. Philadelphia: Saunders College Publishing; 1976. ISBN: 0-03-083993-9, 9780030839931

[57] Marx D, Hutter J. Ab Initio Molecular Dynamics: Basic Theory and Advanced Methods. Cambridge: Cambridge University Press; 2009. ISBN: 978-1-107-66353-4

[58] Shchur LN. Journal of Physics: Conference Series. Vol. 1252. Bristol, United Kingdom: IOP Publishing Ltd; 2019. pp. 012010-012017. DOI: 10.1088/1742-6596/1252/1/012010

[59] Born M, Huang K. Dynamical Theory of Crystal Lattices. New York: Oxford University Press; 1988. ISBN: 978-0198503699

[60] Chen XJ, Zhang C, Meng Y, Zhang RQ, Lin HQ, Struzhkin VV. Mao HK, β-tin→Imma →sh phase transitions of germanium. Physical Review Letters. 2011;**106**:135502-135504. DOI: 10.1103/PhysRevLett.106.135502

[61] Zhang H, Shang SL, Wang Y, Saengdeejing A, Chen LQ, Liu ZK. First-principles calculations of the elastic, phonon and thermodynamic properties of Al12Mg17. Acta Materialia. 2010;**58**: 4012-4018. DOI: 10.1016/j.actamat.2010.03.020

[62] Bian Q, Bose SK, Shukla RC. Vibrational and thermodynamic properties of metals from a model embedded-atom potential. Journal of Physics and Chemistry of Solids. 2008; **69**:168-181. DOI: 10.1016/j.jpcs.2007.08.046

[63] Moser D, Baldissin G, Bull DJ, Riley DJ, Morrison I, Ross DK, et al. The pressure–temperature phase diagram of MgH2 and isotopic substitution. Journal of Physics: Condensed Matter. 2011;**23**: 305403-305800. DOI: 10.1088/0953-8984/23/30/305403

[64] Turney JE, Landry ES, McGaughey AJH, Amon CH. Predicting phonon properties and thermal conductivity from anharmonic lattice dynamics calculations and molecular dynamics simulations. Physical Review B. 2009;**79**:064301. DOI: 10.1103/PhysRevB.79.064301

[65] Kohanoff J. Phonon spectra from short non-thermally equilibrated molecular dynamics simulations. Computational Materials Science. 1994; **2**:221-232. DOI: 10.1016/0927-0256(94)90103-1

[66] Cao L, Stoltz G, Lelièvre T, Marinica MC, Athènes M. Free energy calculations from adaptive molecular dynamics simulations with adiabatic reweighting. The Journal of Chemical Physics. 2014;**140**:104108. DOI: 10.1063/1.4866811

[67] Wang CZ, Chan CT, Ho KM. Tight-binding molecular-dynamics study of phonon anharmonic effects in silicon and diamond. Physical Review B. 1990; **42**:11276-11283. DOI: 10.1103/PhysRevB.42.11276

[68] Debye P. Zur Theorie der spezifischen Wärmen. Annals of Physics. 1912;**39**:789-839. DOI: 10.1002/andp.19123441404

[69] Anderson OL. A simplified method for calculating the Debye temperature from elastic constants. Journal of Physics and Chemistry of Solids. 1963;**24**:909-917. DOI: 10.1016/0022-3697(63)90067-2

[70] Deus P, Schneider HA, Voland U. Estimation of the Debye temperature of

diamond-like semiconducting compounds by means of the Lindemann rule. The Journal Crystal Research and Technology. 1981;**16**:951-948. DOI: 10.1002/crat.19810160814

[71] Wolf U, Bohmhammel K, Wolf G. Supports open access. Thermochim Acta. 1998;**310**:37-42. DOI: 10.1016/S0040-6031(97)00382-1

[72] NIST-JANAF. Thermochemical Tables. Last Update to Data Content. College Park, Maryland, United States DOI: Available from: https://janaf.nist.gov: American Institute of Physics; 1998 [Accessed on 21 November, 2021]

[73] Barril X, Orozco M, Luque FJ. Predicting relative binding free energies of tacrine-Huperzine a hybrids as inhibitors of acetylcholinesterase. Journal of Medicinal Chemistry. 1999;**42**: 5110-5119. DOI: 10.1021/jm990371u

[74] Frenkel D, Smit B. Understanding Molecular Simulation: From Algorithms to Applications. Cambridge, Massachusetts, United States: Academic Press; 2002. DOI: 10.1016/B978-0-12-267351-1.X5000-7 ISBN: 978-0-12-267351-1

[75] Gowdini E, Ahmad AA, Mabudi A, Hadipour NL, Kharazian B. A molecular dynamics study on the thermal properties of carbon-based gold nanoparticles. The Journal of Molecular Modeling. 2020;**26**:307-309. DOI: 10.1007/s00894-020-04559-2

Section 2

Calorimetry in Materials

Chapter 4

Comparative Study of Setting Time and Heat of Hydration Development of Portland Cement According to EN 196-3

Katalin Kopecskó and Attila Baranyi

Abstract

One of the most critical properties of cementitious materials is the initial (IST) and final (FST) setting time, which helps to plan the transportability, workability and demoulding of concrete over time. The standards used to determine the setting time are based on measurement of penetration resistance; these are measured of the depth of penetration with a well-defined body (usually a Vicat needle) into a cement paste as a function of time. Two European standards deal with setting time: EN-196-3 and EN 480-2. EN 196-3 is used to determine the setting time of cement paste of standard consistency. Semi-adiabatic calorimetry (SAC) can be a suitable method for determining the setting time of cementitious materials and concretes of non-standard consistency. This method examines the heat evolution of the hydration reaction of cement. The heat evolution is proportional to the change in viscosity during the setting process and to the Vicat needle penetration depth. This study aimed to find a simple, more accurate and cheaper alternative measurement method for determining the setting time of cementitious materials, which can also be applied to concretes.

Keywords: Portland cement, setting time, IST, FST, semi-adiabatic calorimetry, SAC, EN-196-3

1. Introduction

One of the most important parameters of the cementitious materials is the initial (IST) and final (FST) setting time. Knowing these data, it is possible to plan the maximum workability, casting (IST), and the initial hardening time can be estimated (FST) for formwork removal. The hydrate compounds formed during the hydration of clinker minerals (alite, belit, celite, felite) that evolve a solid matrix from the viscous slurry. The standards used currently measure this transformation process using various penetration resistance procedures (ASTM C191–19, ASTM C266–20, ASTM C403/C403M-16, ASTM C807–20, ASTM C953–17, AASHTO T131–20, AASHTO T154–18, ISO 9597:2008). These technics can be for example the Vicat needle test, Gillmore needles test, and the Hilti nail gun test [1].

The most common method for setting time measurement is the Vicat needle test. This method is applied by EN 196–3. A needle of well-defined weight and diameter is penetrated into cement paste of standard consistency during the process. The penetration of the needle is inversely proportional to the actual viscosity of the test sample, thus and the setting progress.

Despite the simplicity and prevalence of this method, several attempts have been made to induce the penetration method as this technique is applicable only for cement sludge and does not allow monitoring of the whole cement setting process. Such a method includes measuring ultrasonic impulse velocity [2–6] and the electrical resistance [7–9]. These methods are complex in structure and require high-level expertise that makes their use difficult.

Recently, the measurement based on semi-adiabatic calorimetry (SAC) has become popular for determining the setting time of cementitious materials [10–18]. During the procedure, the change in the hydration heat development of the cement is monitored as the clinker minerals chemically react with the water, which results in heat generation (exothermic reaction). The resulting temperature change usually follows the change in mechanical properties as well. The advantage of thermal analysis is the simple design of the measuring system which consists only of a thermally insulated vessel (calorimeter), a thermocouple and a data logger. In addition, ASTM C1679 and ASTM C1753 can help design and improve the successful execution of SAC tests.

During our measurement, we observed several disadvantages of the Vicat method prescribed by the EN 196–3 standard: discontinuous, inaccurate method, the drop number is limited, the examination of non-standard consistency (slightly plastic) mixtures is difficult. The standard measurement of the end of the setting time can be challenging with an automatic device, especially in the case of cementitious pastes with a long setting time [19–21], and automated devices are expensive equipment.

In this study, we compared the setting time of CEM I 42.5 N pure Portland cement according to EN 196–3 with the heat evaluation profile recorded during the calorimetric tests with three different water/cement (w/c) ratios. The measured cement paste was prepared using only deionised water and cement; no admixtures were added for the mixture preparation.

According to our hypothesis, the Vicat method can be substitute heat hydration measurement of the cement paste based on the results obtained from the synchronising of the two processes.

2. Short introduction of EN 196-3 standard

Currently, there are two European standards in force for determining the setting time of cementitious materials. The EN 196–3 cement testing method and the EN 480–2 deal with investigating concrete admixtures, but this standard also uses the Vicat method.

The same method is used for ASTM standards. These methods also measure the penetration resistance during the setting of cementitious materials using the Vicat and the Gillmore method. However, these standards already include mortar and concrete testing (ASTM C403/C403M-16, ASTM C807–20, ASTM C953–17).

The EN 196–3 standard principle is *"cement paste of standard consistency has a specified resistance to penetration by a standard plunger."* The water content of the cement paste of standard consistency must be determined from several different mixtures. The setting time is defined by *"observing the penetration of a needle into cement paste of standard consistency until it reaches a specified value."*

The standard specifies the following parameters:

- The laboratory in which specimens are prepared and tested shall be maintained at a temperature of (20 ± 2) °C and relative humidity of not less than 50%.
- The mixer must be conforming to EN 196–1.
- The procedure of the cement paste: quantity of cement (500 ± 1 g), the method of mixing:
    - the cement and water must be added within 10 s (zero time),
    - start mixing at slow speed (140 ± 5 rpm) for 90 s,
    - the mixing must be stopped for 30 s – during this time the cement paste must be scraped off the wall of the bowl and placed in the middle of it,
    - it follows another 90 s mixing, so the whole mixing time is 30 min.
- The temperature of the cement, the water and the apparatus used for the specimen preparation must be 20 ± 2°C,
- During the underwater tests, the water bath must be thermostatically controlled at (20.0 ± 1.0) °C.
- Vicat apparatus:
    - Plunger for determination of standard consistency: cylinder of at least 45 mm effective length and of (10.00 ± 0.05) mm diameter (**Figure 1**).
    - Vicat needle for initial set: steel and in the form of a right cylinder of effective length of at least 45 mm and diameter (1.13 ± 0.05) mm (**Figure 1**).
    - The total mass of moving parts shall be (300 ± 1) g.
    - Needle with attachment for final set: a needle with a ring attachment of diameter approximately 5 mm (**Figure 1**)
    - Vicat mould: it shall be made of hard rubber, plastics or brass. It shall be of cylindrical or truncated conical form (40.0 ± 0.2) mm deep and shall have an internal diameter of (75 ± 10) mm. A base plate must be placed which is larger than a ring and at least 2.5 mm thick, waterproof and resists to the effect of cement paste.

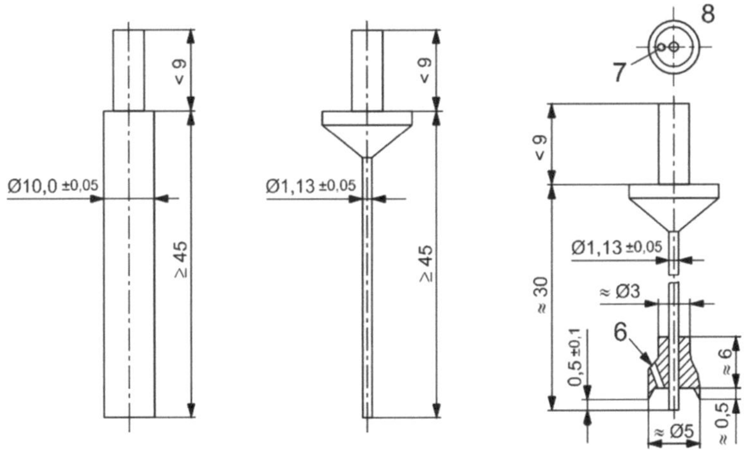

Figure 1.
Plunger for determination of standard consistency (left) Vicat needle for initial set time determination (middle), needle with attachment for final set time evaluation (right) (EN 196–3).

The cement paste must be prepared with a standard mixer according to the method as described in the EN 196–1. Then the lightly oiled Vicat mould is placed at a lightly oiled base plate (usually glass), then filled it with cement paste without compaction or vibration. The standard consistency of cement paste with smoothed upper surfaced is measured; then the plunger is changed for a needle at least 45 mm length and 1.13 $\pm$ 0.05 mm in diameter, and the initial setting time for this water content is determined.

For initial setting time (IST), the standard refers to the time that elapses between the mixing of the cement paste (zero time) and the time until the distance between the needle and the base plate is 6 $\pm$ 3 mm (penetration 34 $\pm$ 3 mm). To determine the final setting time (FST), the mould should be inverted, and the measurement must be continued with the needle with attachment. The time when the needle is already less than 0.5 mm penetrated in the cement paste is considered to be the final setting time, so only the needle and not the attachment leaves a mark on the surface of the specimen.

3. The problems during the standard measurement

The Vicat method, according to EN 196–3 is a penetration measurement, which can only be applied to standard consistency cement paste. It is possible to measure mortar with a modified Vicat test recommended by ASTM C807, but this method cannot be used for mortars with non-standard consistency. For the determination of setting time of concretes, an approximate result can be obtained by penetration resistance measurement using ASTM C403.

The method recommended by EN 196–3 can be challenging to determine for mixtures with non-standard consistency (viscosity). The Vicat mould can easily tip over, and the lower viscosity materials may leak at the bottom of the ring. Therefore, we designed a threaded Vicat mould (**Figure 2**) that also includes the base plate and fits precisely on the rotating plate of the Controls Vicamatic2 instrument used in our studies.

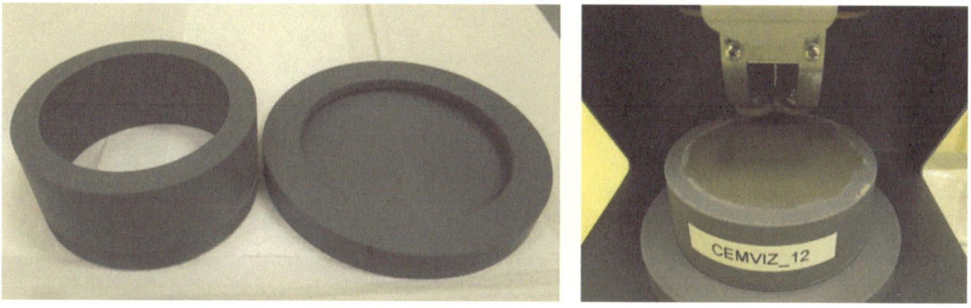

Figure 2.
Threaded Vicat mould.

After determining the IST, the threaded Vicat mould can be disassembled as needed and then inverted with the cement paste placed in it to measure FST. However, in the case of measurements with automatic Vicat instrument, it is impossible or very complicated to determine the end of the setting time according to the standard. Thus, the needle for the initial set is most often used for the whole test, and the sample is usually not reversed. In this case the shrinkage of the specimen must be considered by the effect of which FST seemingly appears at penetration rate more than 0.5 mm in the case of most measurements.

One of the main disadvantages of this method is from its discontinuous nature, due to which we cannot monitor the binding process precisely. The penetration resistance of cement paste which is proportional to changes in the viscosity of the material can be estimated by only individual "sampling".

The automatic device we use can perform 44 standard drops, which means that when measuring mixtures with unknown setting times, this "sampling" has to be managed, especially in the case of cement pastes with a long setting time [19–21]. We can miss the IST if we start the measurement too early or set up a too-long delay time on the automatic Vicat device. In addition, care must be taken to set the proper drop sequence in order to achieve sufficient measurement accuracy.

4. Determination of setting time by thermal method

In order to avoid the problems mentioned above and to be able to monitor the setting process, semi-adiabatic calorimetry (SAC) is an increasingly widespread method for determining the setting time of cementitious mixtures [10–18].

The SAC method is suitable for determining the setting time of cement paste, mortar, and concrete, respectively. We can monitor the setting process; it is possible to measure more accurately. It is not disturbed by the shrinkage of the specimen. It can be done with a simpler and cheaper measuring instrument.

The ASTM C403 standard proposes two methods for the thermal determination of the setting time:

- The **Derivatives method** determines the initial setting time (IST), as the time which results from the maximum curvature of the second derivative of the hydration temperature–time function (**Figure 3**). It defines the final setting time (FST), as the time corresponding to the peak of the first derivative curve of the

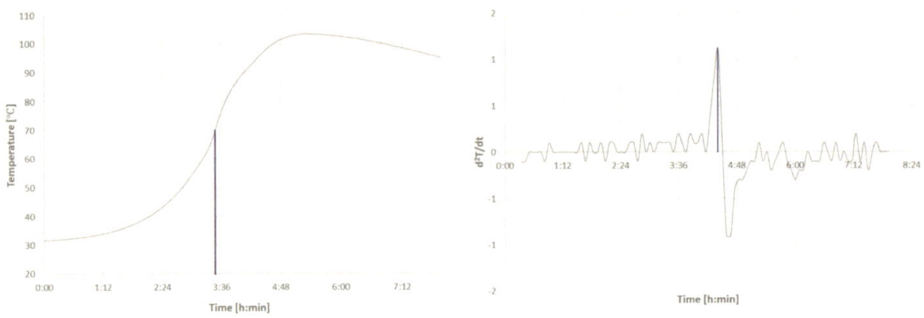

Figure 3.
Determination of IST from the derivatives method.

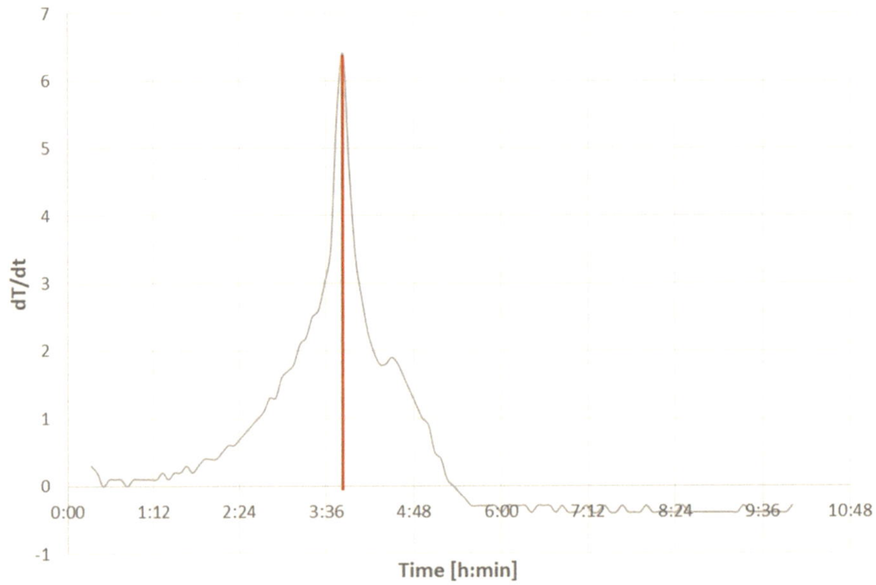

Figure 4.
Determination of FST by the derivatives method.

temperature–time function (**Figure 4**). This method may function well for clear data sets, but it is sensitive to the peaks not belonging to the ones occurring in data and to environmental changes [10, 11, 14, 15, 18].

- The **Fractions method** defines the initial and final setting times as a percentage of the temperature rise below the main hydration peak from baseline to the measured maximum (**Figure 5**). The initial and final setting times are determined as a percentage of the total semi-adiabatic temperature rise of the sample. 21% and 42% are the default initial and final setting time values at standard laboratory curing conditions [10, 11, 14, 18].

This method is more stable than the Derivatives method, but it is more sensitive to the definition of baseline temperature [10].

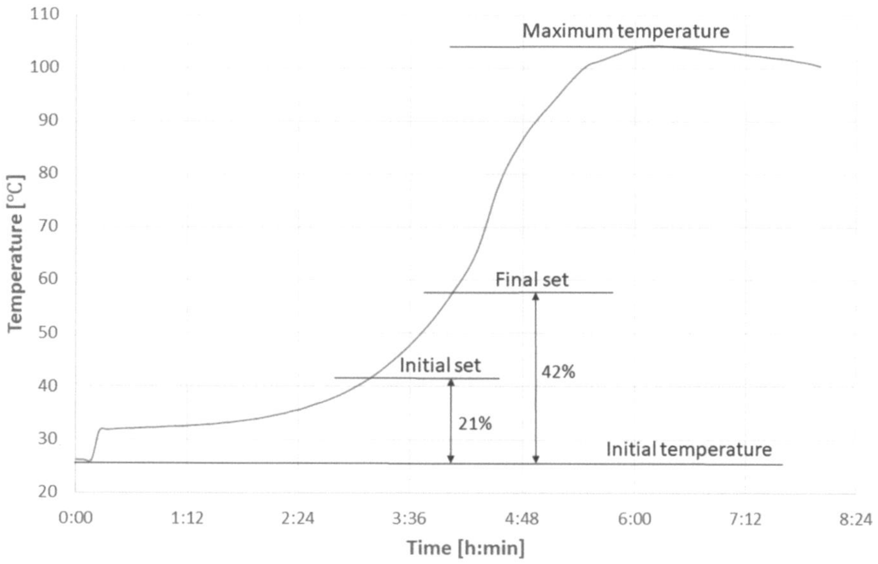

Figure 5.
Determination of the IST and FST as defined by fractions method.

5. The comparative measurement heat of hydration and the setting time according to EN 196-3

In our work, we investigated whether there is a simple correlation between the standard Vicat method and the heat evolution measurement results similar to the ASTM standard methods, which could lead to a method based on in many respects cumbersome and inaccurate penetration measurements. The tests were performed using CEM I 42.5 N cement and deionised water at 0.25, 0.28 and 0.31 water/cement ratios (w/c) at 26 ± 0.5°C. For the measurement we used Controls Vicamatic2 type automatic Vicat device, 300x400x300 mm polystyrene calorimeter and Comet M1200 data logger. We made three parallel measurements from each w/c; then the results were averaged.

In the first step, we synchronised the clock of the Vicat device and the data logger, and then we prepared the cement paste with the given w/c ratio according to the standard. For mixing, a standard Controls 65-L0502 mortar mixer was used to prepare a sufficient amount of sample for both Vicat and heat evolution testing.

For the standard setting time measurement, the threaded Vicat mould (**Figure 2**) was filled with cement paste and placed on the rotating plate of the automatic Vicat device. Air bubbles were removed from the sample by gentle tapping, and then the surface was smoothed.

For the semi-adiabatic (SAC) measurement, 1800 g (about 1 dm^3) of the same cement paste was filled into a 1 l beaker pre-smeared with a form release agent. The sample was placed in the calorimeter. The Teflon-coated thermocouple was then immersed so that it extended to the centre of the sample (**Figure 6**).

During the measurement of heat of hydration the frequency of the temperature data collection was 5 min. By increasing the w/c, the Vicat measurement had to be started with a delay because the binding started later due to the higher water content.

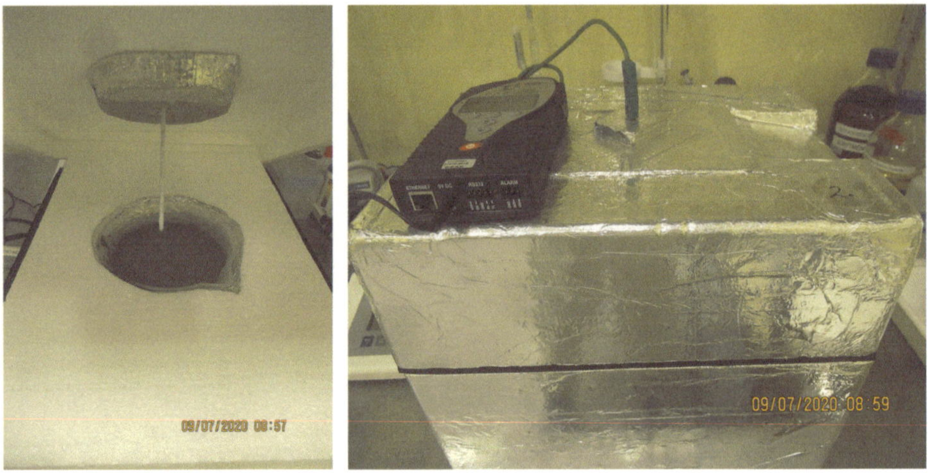

Figure 6.
Polystyrene calorimeter with data logger.

6. Results and discussion

Data from the standard Vicat method and heat evolution measurements were synchronised and plotted on one graph (**Figures 7–9**). We represented the depth of each penetration of Vicat measurement on bar graphs in the setting time – heat development diagrams, while the blue curve shows the hydration heat development of the cement. The penetration at IST is marked with a blue bar; and the FST is marked with a red column according to EN 196–3 standard.

According to the standard, when the penetration depth of the needle sinks only 34 ± 3 mm (6 ± 3 mm from the base plate) into the test specimen is considered to be IST. Thus, the time corresponding to the penetration depths of 37 and 31 mm can also be considered IST, which means that it is up to the person performing the measurement to consider which time he considers the initial setting time. This method can cause a problem, especially when measuring mixtures with long setting times, where this interval can mean a difference of several hours [19–21]. Determining the end of the setting time is even more of a problem, as penetration damages the surface of the specimen during the test and lowers it due to possible shrinkage, which makes it difficult to determine the FST accurately.

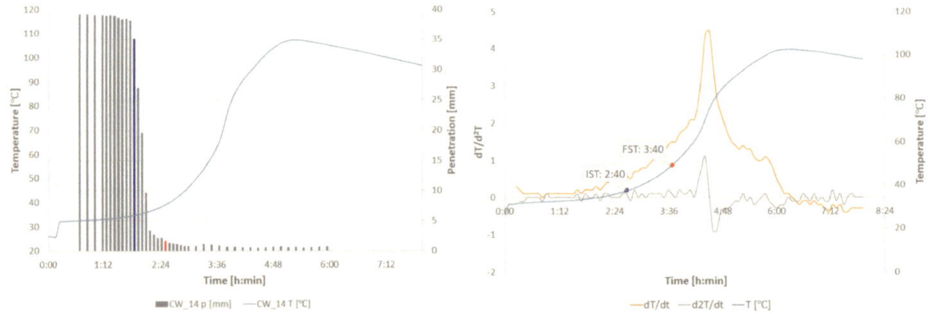

Figure 7.
Setting time—Heat evolution function (left), derivative method (right), w/c = 0.25.

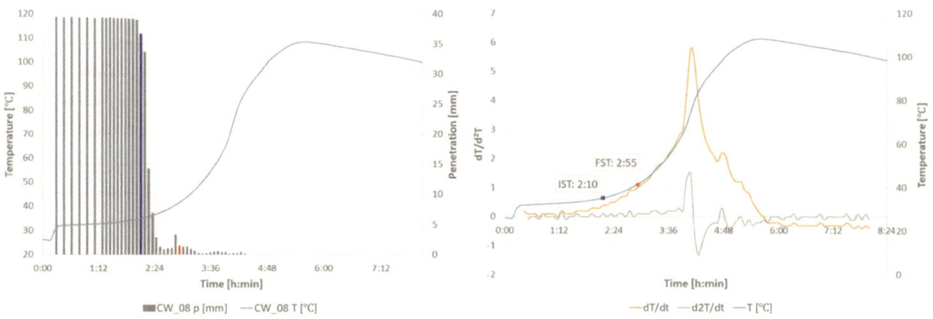

Figure 8.
Setting time—Heat evolution function (left), derivative method (right), w/c = 0.28.

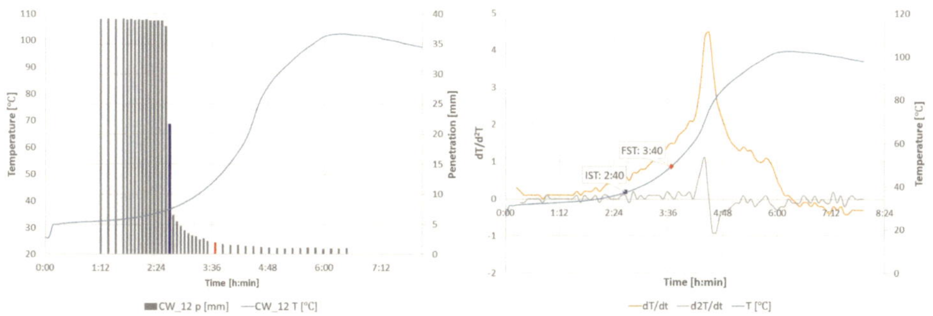

Figure 9.
Setting time—Heat evolution function (left), derivative method (right), w/c = 0.31.

We considered the initial setting time to be the first penetration values smaller than 37.00 mm during our measurements. However, in determining the FST, we compared the previous test performed according to the standard (needle exchange, inversion of the specimen) with the result of the test performed with the needle used for the initial set, without inverting the specimen. Based on these, the time of penetration less than 2.00 mm below the level previously calibrated to 40 mm was considered to be the final setting time.

During the cement paste preparation previously, the temperature suddenly rises, then after a dormancy period, the temperature starts to rise again. In the case of the cement paste tested in the calorimeter, the maximum temperature exceeded 100°C in all cases. It was found that by increasing the w/c factor, the setting time is also extended in proportion to the initial temperature: w/c = 0.25 IST = 1:35–1:50, w/c = 0.28 IST = 2:05–2:15, w/c = 0.31 IST = 2:35–2:45 (see **Table 1**). Slower hydration is likely caused by an increased distance between particles [22]. Another factor may be that more water absorbs more heat, which is not used to accelerate the binding reaction.

At 0.25 w/c ratio, the initial setting time occurred at a temperature rise of 8.3–8.7°C, while at 0.31 w/c ratio, an initial temperature (26 $\pm$ 0.5°C) of up to 10.5°C (**Table 1**). However, the temperature fluctuation of the mixtures prepared with the same w/c ratio measured at IST remained below 0.3°C, therefore the IST and FST values can be determined by measuring the temperature difference (ΔT_{IST}, ΔT_{FST}) similarly to the Fractions method at the same conditions. During our measurements, the initial setting

		Vicat	Vicat-SAC		Vicat	Vicat-SAC		SAC	
Sample name	w/c	IST [h:min]	T_{IST} [°C]	ΔT_{IST} [°C]	FST [h:min]	T_{FST} [°C]	ΔT_{FST} [°C]	T_{max} [°C]	$\Delta t_{Tmax.}$ [h:min]
CW_05	0.25	1:35	34.1	8.3	2:20	38.1	14.9	106.4	5:05
CW_06	0.25	1:40	33.9	8.3	2:40	42.0	16.4	107.7	5:00
CW_07	0.25	1:50	34.7	8.7	2:30	39.1	13.1	107.2	5:15
CW_08	0.28	2:05	35.2	9.1	3:20	49.4	23.3	109.0	5:25
CW_09	0.28	2:05	34.8	8.8	2:55	41.4	15.4	108.2	5:35
CW_10	0.28	2:15	35.4	9.4	3:05	43.3	17.3	107.0	5:35
CW_11	0.31	2:35	36.7	10.5	3:20	43.7	17.5	104.2	6:10
CW_12	0.31	2:40	36.8	10.5	3:40	47.7	21.4	102.5	6:20
CW_13	0.31	2:45	35.9	9.9	4:05	51.7	25.7	103.4	6:30

Table 1.
Results of comparative measurements of Vicat and SAC method.

time occurred at w/c = 0.25 is approx. 8.5°C, at w/c = 0.28 is approx. 9.0°C, and at w/c = 0.31 approx. 10°C temperature rise.

Notations: Vicat—setting time values obtained by the Vicat method, SAC—results of semi-adiabatic calorimetry, Vicat-SAC—temperature values for setting times measured with Vicat needle.

We examined the applicability of the Derivative method, but as shown in **Figures 7–9**, we observed significant differences in the measurement results. Using the Derivative method, IST was 1:40–2:00 delayed compared to the Vicat method, irrespective of w/c, while FST followed it with a 5–10 min lag. Thus, FST occurred with a delay of 0:35 and 1:20 compared to the standard method after the measurement.

Table 1 summarises the measurement results of the Vicat method and heat evolution. IST and FST denote the initial and final setting time, T_{IST}, T_{FST} mark the temperature value measured at this time, and ΔT_{IST} and ΔT_{FST} are the temperature difference measured from the initial temperature (26 ± 0.5°C). The $T_{max.}$ and $\Delta T_{max.}$ are the maximum temperature or the corresponding temperature difference, while $\Delta t_{Tmax.}$ shows the spent time until the maximum temperature value is reached.

We also found similar differences in the Derivatives method with the Fractions method. **Table 2** shows the temperature values ($T_{Fract\ IST}$, $T_{Fract\ FST}$) for the given w/c ratio, calculated by the Fractional method, and the corresponding FSTs (Fract IST, Fract FST), and the difference between the results of the Fractions and the Vicat method ($\Delta t_{Fract\text{-}IST}$, $\Delta t_{Fract\text{-}FST}$) are indicated.

It can be seen that the IST values calculated by the Fractions method give on average 40–65 min longer and the FST 31–56 min longer set values. The proportional factors (k_{IST}, k_{FST}) calculated by us in relation to the setting time measured by the Vicat method, as a function of the w/c ratio, are summarised in **Table 3** (Eq. (1)). It was found that the constant proportionality factors of 21% and 42% given for the Fractions method increase with the w/c ratio.

$$k_{ST}\,[\%] = \frac{\Delta T_{ST}}{\Delta T_B} \bullet 100 \qquad (1)$$

Sample name	w/c	$T_{Fract\ IST}$[°C]	Fract IST [h:min]	$\Delta t_{Fract\text{-}IST}$ [h:min]	$T_{Fract\ FST}$ [°C]	Fract FST [h:min]	$\Delta t_{Fract\text{-}FST}$ [h:min]
CW_05	0.25	42.7	2:45	1:10	59.7	3:25	1:05
CW_06	0.25	42.8	2:45	1:05	60.1	3:25	0:45
CW_07	0.25	43.1	2:50	1:00	60.1	3:30	1:00
CW_08	0.28	43.5	3:00	0:55	60.9	3:45	0:25
CW_09	0.28	43.3	3:05	1:00	60.5	3:45	0:50
CW_10	0.28	42.0	3:05	0:50	60.0	3:50	0:45
CW_11	0.31	42.4	3:15	0:40	58.8	4:05	0:45
CW_12	0.31	42.1	3:15	0:35	58.1	4:10	0:30
CW_13	0.31	42.3	3:30	0:45	58.5	4:25	0:20

Table 2.
Results of comparative measurements of Vicat and fractions method.

w/c	k_{IST} [%]	k_{FST} [%]
0.25	10	18
0.28	11	23
0.31	13	28

Table 3.
The calculated proportional factors compared with the Vicat values.

where ΔT_{ST} is the temperature change belonging to IST or FST; ΔT_B is the difference between the maximum and the initial (room) temperature.

Based on these results the setting times can be calculated as follows (Eqs. (2) and (3)):

$$\Delta T_{IST}\ (IST) = k_{IST}\ (w/c) \bullet \Delta T_B, \tag{2}$$

Sample name	w/c	$\Delta t_{Tmax\text{-}IST}$ [h:min]	$\Delta t_{Tmax\text{-}FST}$ [h:min]
CW_05	0.25	3:30	2:45
CW_06	0.25	3:20	2:20
CW_07	0.25	3:25	2:45
CW_08	0.28	3:20	2:05
CW_09	0.28	3:30	2:40
CW_10	0.28	3:20	2:30
CW_11	0.31	3:35	2:50
CW_12	0.31	3:40	2:40
CW_13	0.31	3:45	2:25

Table 4.
The differences between the time of maximum temperature and Vicat values.

$$\Delta T_{FST}\,(FST) = k_{FST}\,(w/c) \bullet \Delta T_B \quad (3)$$

In our experiments, we assumed that for a given type of cement, with a defined w/c factor, for a given initial temperature and preparation method (see EN 196–3), the heat of reaction (heat of hydration) is the same under semi-adiabatic conditions. This also means that the reaction rate is constant, so the ratio of the time to maximum temperature ($\Delta t_{Tmax.}$) and the setting times (IST, FST) do not change. So, we calculated these differences (Eq. (4)):

$$\Delta t_{Tmax-IST} = \Delta t_{Tmax.} - IST;\ \Delta t_{Tmax-FST} = \Delta t_{Tmax.} - FST \quad (4)$$

, which proved to be nearly constant as expected. The fluctuations in the results were largely due to an error in the Vicat method (**Table 4**).

7. Conclusions

The Vicat method prescribed by the EN-196-3 standard is a widespread measurement that is still in use today mainly due to its simple implementation. The measurement is used to determine the setting time of standard consistency cement paste; however, it is not suitable for non-standard flow cement paste or for testing mortars and concretes. It does not give a comprehensive picture of the setting process; it is inaccurate, which is a problem, especially in the case of cement pastes with a longer setting time. The measurement result depends to a large extent on the skill of the person performing the measurement, and in the case of the use of automatic Vicat devices, the end of the setting time cannot be determined according to the standard, or only with a great difficulty.

The semi-adiabatic calorimetric (SAC) method measures the amount of heat (reaction heat) released during the hydration reaction of a cementitious material, which is proportional to the reaction rate of the clinker minerals (and additives), i.e. the hardening process. The heat of hydration depends on the fineness of the grinding fineness of the cement, the w/c ratio, the method of preparation, the ambient temperature, the quality and quantity of additional materials. This means that for a given cement type and w/c ratio, at a constant environmental temperature, with the same mixing mode, the hydration heat development process takes place in the same way. Thus, for a given mixture, the time to the initial and final setting times and to the maximum of heat evolution is almost constant.

During our measurements, the setting properties of CEM I 42.5 N cement were investigated at 0.25, 0.28, and 0.31 w/c ratios with deionised water without admixtures and aggregates. We sought a parameter comparing the results of the heat evolution curve and the Vicat penetration test to estimate the initial (IST) and final (FST) setting time. As w/c increases, the setting time (ST) values extend and the associated temperature changes also increase, so these characteristics can only be used to determine the setting time for the mixtures of the same composition (Eqs. (5)–(7)).

$$w/c = 0.25 \qquad IST \propto T_0 + 8.5°C \quad (5)$$

$$w/c = 0.28 \qquad IST \propto T_0 + 9.0°C \quad (6)$$

$$w/c = 0.31 \qquad IST \propto T_0 + 10.0°C \quad (7)$$

where T_0 is the initial temperature.

The IST, FST and the time to maximum temperature ($\Delta t_{Tmax.}$) occur proportionally later for mixtures with higher w/c ratios, however, the time between them ($\Delta t_{Tmax\text{-}IST}$, $\Delta t_{Tmax\text{-}FST}$) is almost constant (**Table 3**).

For the mixtures used in our studies, the IST and FST were measured 3.5 and 2.5 hours before Δt_{Tmax}. Minor fluctuations were observed only in the determination of FST values caused by the uncertainty resulting from the Vicat method. We can conclude that the setting time (ST) of the cement paste made of ordinary Portland cement without admixtures and aggregates can be determined without the use of Vicat apparatus. It can be determined accurately by SAC method with the knowledge of the time to maximum temperature (Δt_{Tmax}) corresponding to the maximum temperature value ($T_{max.}$) (Eqs. (8) and (9)):

$$IST \propto \Delta t_{Tmax} - 3:30 \qquad (8)$$

$$FST \propto \Delta t_{Tmax} - 2:30 \qquad (9)$$

Therefore, the SAC method is well applicable to simple Portland cement-water mixtures in the range of 0.25–0.31 w/c. Using too much mixing water (w/c > 0.44) will cause cement paste bleeding (water separation from the cement). Excess water does not participate in the chemical reaction of setting, but at the same time, it impedes the setting process and, due to its high heat capacity, draws heat out of the system.

The applicability of the SAC method to the investigation of the setting properties of CEM I 42.5 N Portland cement was confirmed at 0.25, 0.28, and 0.31 w/c ratios. Other cement pastes or mortars and concretes prepared without admixtures and supplementary materials may behave similarly, but it needs to be verified by further testing.

Acknowledgements

The authors of this article would like to thank Dr. László Csetényi for his helpful comments and supports, as well as Dr. Tamás Kói for his help in the mathematical problems occurred.

The authors acknowledge the support by the Hungarian Research Grant NVKP_16-1-0019 "Development of Concrete Products with Improved Resistance to Chemical Corrosion, Fire or Freeze-Thaw."

List of referred standards

EN 196–3: 2017 Methods of testing cement. Part 3: Determination of setting time and soundness.

EN 480–2:2007 Admixtures for concrete, mortar and grout. Test methods. Part 2: Determination of setting time.

ASTM C191–19 Standard Test Methods for Time of Setting of Hydraulic Cement by Vicat Needle. ASTM International. West Conshohocken. PA. 2019. DOI: 10.1520/C0191-19

ASTM C266-20 Standard Test Method for Time of Setting of Hydraulic-Cement Paste by Gillmore Needles. ASTM International. West Conshohocken. PA. 2020. DOI: 10.1520/C0266-20

ASTM C403 / C403M-16. Standard Test Method for Time of Setting of Concrete Mixtures by Penetration Resistance. ASTM International. West Conshohocken. PA. 2016. DOI: 10.1520/C0403_C0403M-16

ASTM C807–20. Standard Test Method for Time of Setting of Hydraulic Cement Mortar by Modified Vicat Needle. ASTM International. West Conshohocken. PA. 2020. DOI: 10.1520/C0807-20

ASTM C953-17. Standard Test Method for Time of Setting of Grouts for Preplaced-Aggregate Concrete in the Laboratory. ASTM International. West Conshohocken. PA. 2017. DOI: 10.1520/C0953-17

ASTM C1679-17. Standard Practice for Measuring Hydration Kinetics of Hydraulic Cementitious Mixtures Using Isothermal Calorimetry. ASTM International. West Conshohocken. PA. 2017. DOI: 10.1520/C1679-17

ASTM C1753 / C1753M-15e1. Standard Practice for Evaluating Early Hydration of Hydraulic Cementitious Mixtures Using Thermal Measurements. ASTM International. West Conshohocken. PA. 2015. DOI: 10.1520/C1753_C1753M-15E01

AASHTO T131–20 Standard Method of Test for Time of Setting of Hydraulic Cement by Vicat Needle. American Association of State and Highway Transportation Officials. 2020.

AASHTO T154–18 Standard Method of Test for Time of Setting of Hydraulic Cement Paste by Gillmore. Needles. American Association of State and Highway Transportation Officials. 2018.

ISO 9597:2008 Cement-test Methods – Determination of Setting Time and Soundness. International Organization for Standardization

Author details

Katalin Kopecskó* and Attila Baranyi
Faculty of Civil Engineering, Budapest University of Technology and Economics, Budapest, Hungary

*Address all correspondence to: kopecsko.katalin@emk.bme.hu

References

[1] Lootens D, Roussel N. Rheology of penetrations tests II: Penetrometers, Vicat and Hilti needles. In: Proceedings of the 12th International Congress on the Chemistry of Cement. Montreal, Canada: National Research Council of Canada; 8-13 July 2007. pp. 1-12

[2] Reinhardt HW, Grosse CU, Herb AT. Ultrasonic monitoring of setting and hardening of cement mortar—A new device. Materials and Structures. 2000; **33**:580-583. DOI: 10.1007/BF02480539

[3] Lee HK, Lee KM, Kim YH, Yim H, Bae DB. Mint: Ultrasonic in-situ monitoring of setting process of high-performance concrete. Cement and Concrete Research. 2004;**34**:631-640. DOI: 10.1016/j.cemconres.2003.10.012

[4] Sant G, Dehadrai M, Bentz D, Lura P, Ferraris CF, Bullard JW, et al. Mint: Detecting the fluid-to-solid transition in cement pastes: Comparison of experimental and numerical techniques. Concrete International. 2009;**236**:53-58

[5] Gabrijel I, Mikulić D, Milovanović B. Mint: Application of ultrasonic measurements for determination of setting and hardening in cement paste. Journal of Civil Engineering and Architecture. 2011;**5**(3):278-283. DOI: 10.17265/1934-7359/2011.03.010

[6] Taylor P, Wang K, Wang X, Wang X. Mint: Comparison of setting time measured using ultrasonic wave propagation with saw-cutting times on pavements. InTrans Project Reports. 2015;**142**:1-30. DOI: 10.13140/RG.2.1.3907.4329

[7] McCarter WJ, Chrisp TM, Starrs G, Blewett J. Mint: Characterization and monitoring of cement-based systems using intrinsic electrical property measurements. Cement and Concrete Research. 2003;**33**:197-206. DOI: 10.1016/S0008-8846(02)00824-4

[8] Wei X, Li Z. Study on hydration of Portland cement with fly ash using electrical measurement. Mint: Materials and Structures. 2005;**38**:411-417. DOI: 10.1007/BF02479309

[9] Li Z, Xiao L, Wei X. Mint: Determination of concrete setting time using electrical resistivity measurement. Journal of Materials in Civil Engineering. 2007;**19**:423-427. DOI: 10.1061/(ASCE)0899-1561(2007)19:5(423)

[10] Wang K, Ge Z, Grove J, Ruiz JM, Rasmussen R, Ferragut T. Mint: Developing a simple and rapid test for monitoring the heat evolution of concrete mixtures for both laboratory and field applications. InTrans Project Reports. 2007;**153**. DOI: lib.dr.iastate.edu/intrans_reports/153

[11] Cost VT, Gardiner A. Practical Concrete Mixture Evaluation via Semiadiabatic Calorimetry. In: Proceedings of the Concrete Technology Forum 2009, National Ready Mixed Concrete Association. Cincinnati, Ohio, USA: 13-15 May 2009

[12] Ge Z, Wang K, Sandberg PJ, Ruiz JM. Mint: Characterization and performance prediction of cement-based materials using a simple isothermal calorimeter. Journal of Advanced Concrete Technology. 2009;**7**(3):355-366. DOI: 10.3151/jact.7.355

[13] Bentz DP. Mint: Critical observations for the evaluation of cement hydration models. International Journal of Advances in Engineering Sciences and Applied Mathematics. 2010;**2**:75-82. DOI: 10.1007/s12572-010-0017-4

[14] Rolo L. Monitoring of the Cement Hydration Behavior and Determination of Non-standard Laboratory Indicators of Setting Time [Thesis]. Canales y Puertos: Ingeniería de Caminos; 2013. DOI: 10.13140/2.1.1963.6480

[15] Hu J, Ge Z, Wang K. Mint: Influence of cement fineness and water-to-cement ratio on mortar early-age heat of hydration and set times. Construction and Building Materials. 2014;**50**:657-663. DOI: 10.1016/j.conbuildmat.2013.10.011

[16] Chung C, Kim JH, Lee S. Mint: The use of semi-adiabatic calorimetry for hydration studies of cement paste. Journal of the Korea Institute of Building Construction. 2016;**16**(2):185-192. DOI: 10.5345/JKIBC.2016.16.2.185

[17] Sanderson RA, Cann GM, Provis JL. Comparison of calorimetric methods for the assessment of slag cement hydration. Advances in Applied Ceramics. 2017;**116**: 186-192. DOI: 10.1080/17436753.2017.1288371

[18] Kang X, Lei H, Xia Z. Mint: A comparative study of modified fall cone method and semi-adiabatic calorimetry for measurement of setting time of cement based materials. Construction and Building Materials. 2020;**248**. DOI: 10.1016/j.conbuildmat.2020.118634

[19] Egan PJ. Mint: A comparison between semi-isothermal and semi-adiabatic calorimetry of retarded cement mixes. Advances in Cement Research. 1988;**1**(2):112-118. DOI: 10.1680/adcr.1988.1.2.112

[20] Csetényi LJ. Stability of Borate-Containing Wastes Encapsulated in Cement [PhD Thesis]. United Kingdom: University of Aberdeen; 1993

[21] Han MC, Han CG. Mint: Use of maturity methods to estimate the setting time of concrete containing super retarding agents. Cement & Concrete Composites. 2010;**32**:164-172. DOI: 10.1016/j.cemconcomp.2009.11.008

[22] Bentz DP, Peltz MA, Winpigler J. Mint: Early-age properties of cement-based materials. II: Influence of water-to-cement ratio. Journal of Materials in Civil Engineering. 2009;**21**:512-517. DOI: 10.1061/(ASCE)0899-1561(2009)21:9(512)

Chapter 5

Calorimetry to Understand Structural Relaxation in Chalcogenide Glasses

Balbir Singh Patial

Abstract

Thermal behavior of chalcogen additive materials synthesized via melt quench method can be studied by reheating the bulk samples in differential scanning calorimetry (DSC) or differential thermal analyzer (DTA) experiment. It involves kinetics of structural transformations as three basic characteristic phenomena correspond to glass transition, crystallization and melting are involved. Thermal stability and glass forming ability are important factors from technological point of view in various applications. Thermal stability of glasses can be ascertained based on calorimetric measurements. In the glass transition region (first region in reheating experiment), structural relaxation takes place. The temperature in glass transition region, its heating rate dependence and empirical approaches for estimation of apparent activation energy are useful to determine utility of these materials in various applications.

Keywords: chalcogenide glasses, DSC or DTA, non-isothermal, structural relaxation, glass transition, activation energy

1. Introduction

Long back ago, it was believed that amorphous solids could not be semiconductors as known from the fact that that the existence of semiconductors is connected with the quantum theory of the solids, which in turn is based on the presence of long range order. The interest in amorphous materials grew in 1950's when it was found that the non-crystalline solids and liquids that do not have any periodic structure also behave as semiconductors. Since then, extensive research initiated to explore these materials. These materials have great potential to serve as raw materials for the fabrication of electronic devices and the time is not far off when crystalline materials may be completely replaced by amorphous materials in electronic industries. It is because of the fact that the preparation technique is comparatively cheaper, processes involved are comparatively easier and the devices are of good quality in amorphous field. Amorphous materials can be regarded as amorphous semiconductors, if the energy band gap lies between 0.1 eV and 3 eV. Amorphous materials are becoming progressively popular due to their wide range of applications in solid state devices.

Glasses are those amorphous solids which are prepared by rapid cooling of the melt. In principle, any substance can be made into a glass by cooling it from the liquid state fast enough to prevent crystallization. In general, glass scientists regard the term 'glass' as covering 'all non-crystalline solids that show a glass transition' irrespective of their preparation methods. In actual practice, glass formation has been achieved with a relatively limited number of substances. The two standard methods of preparing amorphous solids are: (*i*) by condensation from the vapor as in thermal evaporation, sputtering, glow-discharge decomposition of gas or other methods of deposition and (*ii*) by cooling from a melt. The first method produces thin films, while the second method provides bulk materials. Materials that are obtained by cooling from the melt are called glasses and generally have a lesser tendency to crystallize as compared with those that can be prepared only by deposition [1]. The crystallization of an amorphous material proceeds by the processes of nucleation and growth. Glass formation is inhibition of crystallization. Glass formation becomes more probable at higher cooling rate, the smaller the sample volume and slower the crystallization rate. Whether a particular liquid can be cooled to form a glass or not, will clearly depend on the rates of the atomic or molecular transport processes involved in the nucleation and crystal growth. Liquids with small crystallization kinetic constants can thus form glasses directly from the melt. This is the basic reason as to why selenium (Se) is a good glass former in contrast to pure tellurium (Te). Silicon and germanium cannot be quenched into the glassy state even at very fast quench rates. When it was proposed that chalcogenide glasses containing large proportions of one or more chalcogen elements namely sulfur (S), selenium (Se) and tellurium (Te) instead of the conventional sixth group element oxygen (O) can act as semiconductors and continuous research in this field is going on. Moreover, the possibility of doped chalcogenide glasses in the well controlled manner has opened up many new directions for the application of chalcogenide glasses in different fields.

The alloys of chalcogenide glasses have a certain range of atomic percentages of each constituent in which they can be glassy and beyond this range they are semi-crystalline or crystalline. From literature, it can be inferred that small samples say binary, ternary or multicomponent compositions were prepared, their nature (crystalline or amorphous) were determined and consequently glass formation region is mapped out for each composition. The range of glass forming region is to some extent reliant on the quench rate or the quantity of material used during the sample preparation. Chalcogenide glasses are prepared by quenching of the liquid below melting temperature. Upon cooling a liquid below its freezing temperature, it will either crystallize or to form a glass. Glasses of different compositions have different regions of glass formation directed by the type of bonds between the constituent elements. An increased tendency to glass formation is possessed by chalcogenide compounds and alloys with predominantly covalent chemical bonds. However, the specific composition upto which glass formation in binary, ternary or multi-component system is possible, cannot be predicted in priori and has to be determined experimentally. For example, in Se-Te system, glass formation is possible upto tellurium content 30 at % [2].

When glass forming liquid is cooled, some of its properties change sharply in a narrow temperature range [3, 4]. Figure shows volume versus temperature plot during the course of temperature lowering experiment (melt quenching). Continued cooling decreases the liquid volume in a continuous fashion and the slope of the smooth volume-temperature curve defining the liquid's volume coefficient of thermal expansion. Eventually, when the temperature is brought low enough, a liquid–solid transition takes place. Its signature in terms of low value of the expansion coefficient i.e. a smaller slope, characterizes a solid. A liquid may solidify in two ways; first one corresponds to route 1 (**Figure 1**) discontinuously to a crystalline solid and

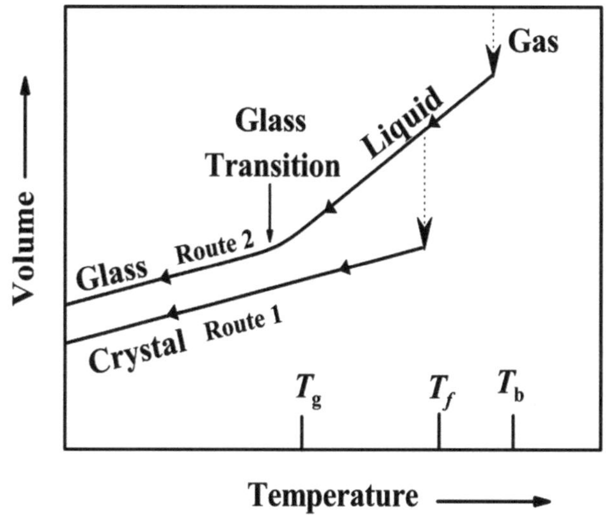

Figure 1.
Volume versus temperature plot for cooling an assembly of atoms can condense into the solid state.

secondly via route 2 continuously to an amorphous solid (glass) (where T_f is freezing temperature and T_b is boiling temperature in **Figure 1**). The liquid-crystal transition occurred at freezing or melting temperature and an abrupt contraction is observed to the volume of the crystalline solid. However, at sufficiently high cooling rates in a melt-quenching experiment, the most materials are found to alter their behavior and follow route 2 to the solid phase. An abrupt change in the slope of the curve equal to the coefficient of volume expansion occurs at a certain temperature. This temperature is called the glass transition temperature (T_g). The liquid-glass transition occurs in a narrow temperature interval near T_g.

There is no volume discontinuity, instead curve bends over to acquire the small slope characteristic of the low thermal expansion of a solid. It is significant that viscosity does not show an abrupt change at the glass transition temperature.

2. Experimental methods

2.1 Melt-quench technique

Melt-quenching technique is a widely used method to prepare the various samples of the chalcogenide glasses. The entire equipment used for preparing the alloys consists of a furnace to prepare the melt of solid mixture, a rocking arrangement to make the melt homogenous and a quenching arrangement to rapidly cooling of the melt. The furnace is operated maximum at 120 V ac and can raise the temperature up to 1200°C or more depending on type of furnace. Quartz ampoules (outer diameter ~ 1.2 cm and inner diameter ~ 1.0 cm, length ~ 10 cm) are prepared by sealing the quartz tube on one side and a neck on other side to pour the solid mixture before evacuation. To prepare glassy alloy of type $Se_{\alpha}In_{\beta}Pb_{\gamma}Te_{\delta}$, the elemental substances (5 N) are weighed according to the exact proportions of high purity (99.999%) elements in accordance with their atomic percentages (α,β,γ,δ are the respective atomic weight percentages of Se, Te, Pb and In).

The quartz ampoules are properly cleaned with soap solution, acetone, methanol and then dried by heating in the furnace at 500°C for half an hour. The weighed mixture is poured into the ampoules and these ampoules are sealed in a vacuum of ~2 × 10^{-5} mbar using the diffusion pump. Then ampoules are fastened to a ceramic rod and the rod is attached to the rocking arrangement. The sealed ampoules are heated in increasing order of melting points of the constituent's elements of the glassy alloys for 2 hours each. For example, in case of Se-Te-In, the temperature of the furnace is raised to 200°C for 2 hours, so that the indium [melting temperature, T_m = 156.6°C] diffuse into the rest of the constituents. The temperature is then raised to 300°C so that the selenium (Se) [T_m = 217°C] thoroughly mixes with other constituents. The sealed ampoules heated upto temperature 600°C in the rocking furnace initially, so that tellurium [T_m = 449°C] in respective composition mixed properly. Later, the temperature of the furnace has been raised upto 900°C and maintained at this temperature for 24 hours and rocking is done to ensure proper mixing and homogeneity of the samples. The heated ampoules are then quickly quenched in ice cooled water to get glassy alloys.

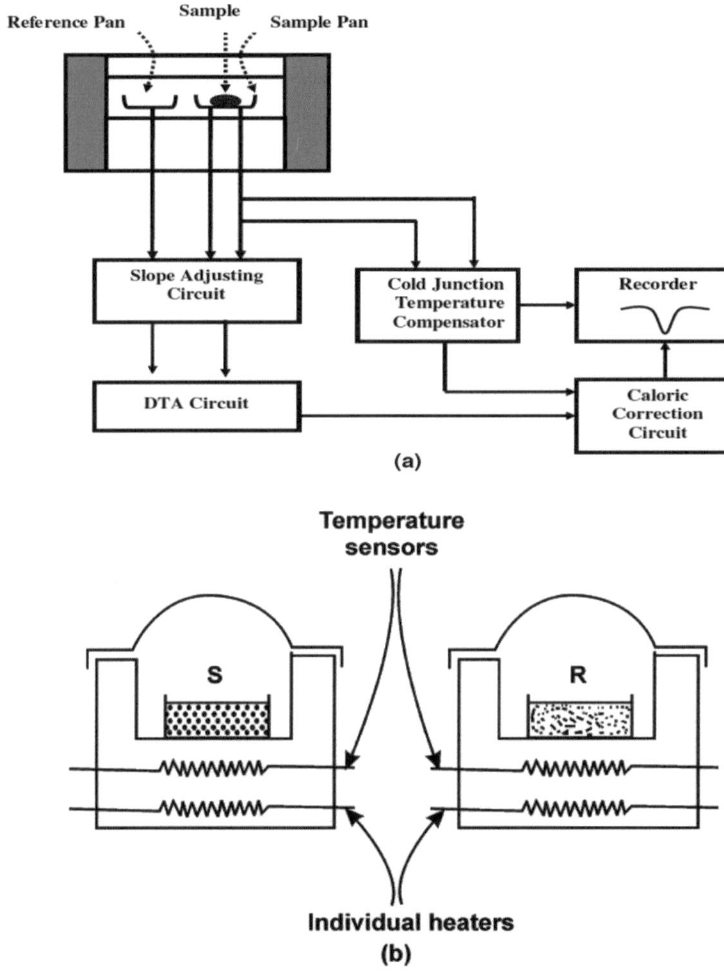

Figure 2.
(a) Block diagram of working principle of DSC. (b) Heating arrangement in DSC.

Then tubes are broken, crushed, separated, ground to fine powder, labeled and kept in vacuum desiccators for further analysis.

2.2 Differential scanning calorimetry

The thermal behavior of the glasses is investigated using DSC system. The complete DSC system consists of the DSC analysis module, temperature controller, data analyzer and recorder/plotter for the preparation of hardcopy record. Working principle of DSC is shown in **Figure 2**. The temperature range covered in this DSC unit was from room temperature to high enough temperature (>500 degree Celsius) as per instrument make. Approximately, 3–5 mg of sample in powder form is encapsulated in standard aluminum pan in an atmosphere of dry nitrogen at a flow of 40 mL min^{-1} and heated as per requirement under either isothermal or non-isothermal conditions. The values of glass transition temperature (T_g), the temperature of crystallization (T_c), and the melting temperature (T_m) are determined by using the microprocessor of the thermal analyzer.

3. Structural relaxation in chalcogenide glasses

Kauzmann [3] showed that glass transition occurs for many types of glasses. It has been suggested that some amorphous solids do not show a glass transition temperature (T_g), but this result has not been established. Since 2005, when I started research in the field of chalcogenide glasses for Ph.D. degree and synthesized binary, ternary and quaternary chalcogen based glasses and even after while continuing research almost 50 samples are prepared by melt quenching technique. But these samples always found to associate with T_g [5–9]. It is because of the fact that rearrangements in the glass structure occur during temperature lowering experiment (cooling) when the glass has the properties of liquid. Therefore, the changes are slow when the glass structure is frozen (cooling is fast enough) and it behaves like a solid. When a glass is cooled rapidly from a temperature, above the transition region into this region, it retains some properties of higher temperature. These properties 'relax' to the characteristics of the lower temperature with time may be pronounced more commonly as '*structural relaxation*'. The time to reach a stable or equilibrium state is longer at lower holding temperatures. If a glass material is cooled rapidly from a room temperature above the transition region, non-uniformities in its temperature during cooling lead to stresses in the glass. These stresses can weaken the glassy material and change its properties. So, it is desirable to remove them by heating the glass at an appropriate temperature in the transition region. At this annealing temperature, the stresses are removed as the glass relaxes. The rate at which annealing goes on is important in preparing glasses for use. The relaxation of glass structure from one set of properties to another after a rapid change in temperature in the glass transition region is called *structural relaxation*. When a glass is subjected to a stress (strain) in this region, the deformation changes with time and this process is called stress or strain relaxation. The rates of this relaxation are of practical importance and have been studied intensively in the last few years.

Figure 3 shows a typical DSC mostly observed in reheating calorimetric experiment displaying glass transition (first endothermic peak), crystallization (exothermic peak) and melting (second endothermic peak) for a chalcogenide glass. But it is pertinent to mention here that heat flow in calorimetric measurements may

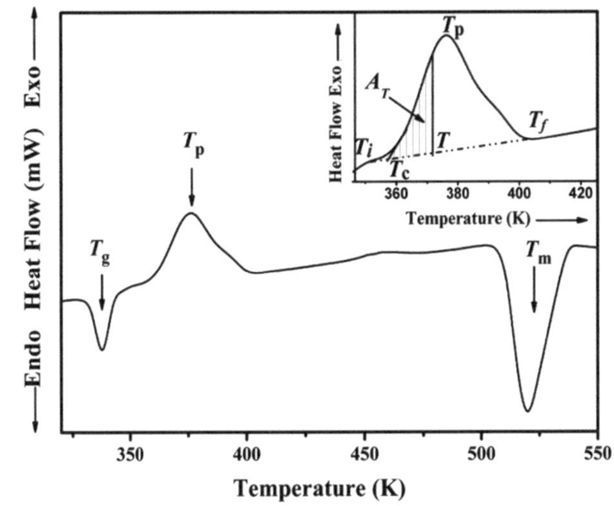

Figure 3.
A typical DSC curve showing glass transition, crystallization and melting of a chalcogenide glass.

be different for other materials like polymers depends on materials' characteristics. The inset of **Figure 2** shows the crystallization fraction at a generic temperature T (shaded area of exothermic peak). DSC curves are characterized by first occurrence of endothermic peak which corresponds to glass transition. In this process, molar volume and enthalpy of chalcogenide material change resulting in variation of specific heat and viscosity which marks transition from solid phase to super cooled liquid phase. In other words, instantaneous change in temperature during quenching causes chalcogenide glassy material to relax from a higher enthalpy to an equilibrium state of lower enthalpy and this process is called thermal relaxation. Glass transition temperature (T_g) corresponds to point of intersection of tangents drawn to baseline and endothermic baseline shift. The onset values of characteristic temperatures are taken while defining the properties and henceforth use of glasses in technological applications. But, for in observations and hence particularly in calculations peak values of T_g, T_p and T_m are taken instead of onset values because of more accuracy in measurement of peak values than onset values. Least-squares fitting method is applied to deduce apparent activation energy and other kinetic parameters while using empirical approaches. Chalcogenide glasses generally show single T_g and single T_p indicate that the investigated system exists in single phase and homogeneous, however, a second glass transition temperature (T_{g2}) is also observed for Se-Te-Sn investigated samples see reference [6]. The existence of a second glass transition temperature directs unusual phase separation happening in thermal treatment. This phase separation, leading to presence of a second or more than two glass transition temperatures, may be thought of arising after super cooled melt transition or the glass may be in two phases or higher to start with itself. The phenomenon of presence of a second T_g has been found in other chalcogenides [10, 11] also.

It is observed that characteristic temperatures shift to higher temperature with increasing heating rate. At T_g, structural relaxation time becomes equal to the relaxation time of observation τ_{ob} [12]. Since $T_g \propto 1/\tau_{ob}$. Thus, with the increase in heating rate, τ_{ob} decreases leads to increased T_g also observed in other glasses [5–9]. It can also be ascribed from heat dissipated so much easier at higher heating rate; therefore, decomposition begins on relatively higher temperature and has high heat of fusion.

T_c is also found to increase (similar to T_g) with increasing heating rate. This could be because the materials do not get enough time for nucleation and crystallization with higher heating rate.

4. Glass transition calorimetry

In reheating experiment of glass in calorimetry, two ways can be adopted namely isothermal and non-isothermal methods. In first method i.e. under isothermal conditions, the material under investigation is placed rapidly to a temperature above T_g and the heat evolved at a constant temperature is noted as a function of time during the crystallization process. In the other method, i.e. under non-isothermal conditions, the material is heated at a fixed heating rate usually from room temperature. Again in this method also, heat evolved is noted as a function of time or temperature. A drawback of experiment under isothermal conditions is the impossibility of attainment a test temperature instantly and during the time, which system desires to stabilize, no observations are possible. However, the second method i.e. under non-isothermal conditions does not have this limitation. Due to above mentioned reasons; we have also applied this technique for the overall crystallization kinetics of investigated glassy alloys [5–9].

Characteristic temperature above which glassy material can have several structural arrangements and below which material is frozen in a structure which cannot change into other configuration easily is defined as glass transition temperature T_g. T_g is an indispensable parameter in studying stability of glassy materials. Furthermore, cohesive forces must be overcome for movement of atoms in glass network. So, T_g must be related to the magnitude of these forces. Therefore, theoretical models are also proposed to determine T_g which assume it to be proportional to cohesive forces, (in terms of mean bond energy $< E >$) and network rigidity. Tichy and Ticha [13] proposed that T_g depend on two factors; coordination number $< r >$ and $< E >$ after analyzing for186 glasses; $T_g = 311[< E > - 0.9]$ (T_g so derived is in kelvin). Lankhorst [14] also devised a model for estimation of T_g from heat of atomization H_s of system and suggested empirical relation; $T_g = 3.44\,H_s - 480$. Tanaka also gave the exponential relationship as [15]; $T_g = exp(1.6 < r > + 2.3)$. Experimental values are sometimes closely matched with theoretically values.

Glass transition temperature depends upon the co-ordination number, bond energy and types of structural units formed. The glass transition region may be studied from glass transition temperature and its heat rate dependence, consequently, apparent activation energy of glass transition. Glass transition temperature represents the strength or rigidity of the glass structure of the glassy material. We employed three approaches to analyze the dependence of the T_g on the heating rate and estimation of apparent glass transition activation energy of bulk glasses [5–9]. The first one corresponds to the empirical relation given by Lasocka [16] as $T_g = A + B\ln(\alpha)$, where A and B are constants depending upon the glass composition. From this relation, it can be easily inferred that a plot between $\ln(\alpha)$ and T_g should be a straight line. Also, A could be deduced as T_g at a heating rate of 1 $Kmin^{-1}$. It is suggested by other researchers that B (determined from slope) may be correlated to the cooling rate of the melt [17, 18]. Lesser is the cooling rate of the melt, lesser is B. Thus, the physical significance of B looks to be connected with the response of the fluctuations in the configuration within the glass transformation region.

Reheating of a chalcogenide glass result into crystallization phenomenon and studies of crystallization kinetics are always connected with the concept of activation energy. Another parameter which is comprehensively used to get into structural relaxation kinetics is apparent activation energy. It is said to be that energy which is absorbed by a assembly of atoms in this region, such that a jump from one metastable state to another is possible. This apparent activation energy is involved in the molecular motion and rearrangement of atoms around glass transition temperature. The more general method used in this regard is Kissinger's equation which is basically meant for the determination of activation energy for crystallization process. The justification of applying this method for the evaluation of the glass transition activation energy comes from the shifting of glass transition peaks at different heating rates similar to crystallization peaks. The details of this method
can be noted from [19, 20] and in case of glass transition it is modified to $(\alpha/T_g^2) = - E_g/ RT_g + constant$. Slope of the plot between $\ln(\alpha/T_g^2)$ and $1/T_g$ is utilized to derive E_g. We have also used this method extensively to deduce E_g, establish a relation with other experimentally deduced parameters and thereafter to draw conclusions regarding thermal stability among the investigated glasses [5–9].

The other approach, using heating rate dependence of T_g is also used defined by Moynihan *et al* [21] in terms of the thermal relaxation phenomenon. In this kinetic interpretation, the enthalpy at a particular temperature and time $H(T, t)$ of the glassy system, after an instantaneous isobaric change in temperature, relaxes isothermally toward a new equilibrium value $H_e(T)$. The relaxation equation can be written in the following form [22]:

$$\left(\frac{\delta H}{\delta t}\right)_T = -\frac{(H - H_e)}{\tau} \tag{1}$$

where τ is a temperature—dependent structural relaxation time and is given by the following relation:

$$\tau = \tau_o \exp\left(-E_g / RT\right)\exp\left[-c\left(H - H_e\right)\right] \tag{2}$$

where τ_0 and c are constants and E_g is the activation energy of relaxation time. Using the above equations, it can be shown [21–23] that $\ln(\alpha) = - E_g/RT_g + constant$. Similarly, in this method i.e. using Moynihan's relation, slope of the linear fit variation of $\ln(\alpha)$ against $1/T_g$ gives E_g. E_g values so deduced are found in concordance with Kisinger's relation [5–9]; therefore, one can use either of these two approaches.

Thermal stability and the ease of glass formation is a major issue in the study of glassy materials as it determines the degree of utilizing the investigated materials in various applications. It can be ascertained based on calorimetric observations. Kauzmann proposed two-third rule to determine the ease in glass formation [3] as $T_{rg} = T_g/T_m$. The composition obeying two-third rule ($T_{rg} \geq 0.65$), indicating that the glass forming ability (*GFA*) for that composition of the material is high.

For a memory and switching material, the thermal stability and *GFA* are of vital importance. Glass transition temperature also gives the worthwhile information about the thermal stability related with strength and rigidity of glass structure. GFA is also associated with cooling of the melt bypassing crystallization. It has been stated that (T_c-T_g) is also an indicator of GFA. Higher the value of this difference, greater is GFA because higher values of this difference indicate the more kinetic resistance to crystallization. One more parameter, Hruby number K_{gl} is important by which thermal stability and glass formation is evaluated as $(T_c\text{-}T_g)/(T_m\text{-}T_c)$ [24]. Higher $(T_c\text{-}T_g)$ delays nucleation while lower $(T_m\text{-}T_c)$ retards growth in nucleated crystals. Thus, Hruby's parameter merges nucleation and growth information during amorphous-crystallization phase transformation. Thus, Hruby's parameter combines the nucleation and growth aspects of phase transformation. Therefore, the composition of the glass with higher value could be taken as the most stable among studied samples.

There is no absolute measurements to define the glass formation, the empirical methods extensively used for its quantifiable properties. The fragility index (F_i) is a significant parameter to define glass forming ability and it is a measure of the rate at which the relaxation time decreases with the increase in temperature around glass transition temperature. This distict parameter is given by the relation [25, 26]; $F_i = E_g/RT_g ln(\alpha)$. According to Vilgis [27], the glass forming liquids that show an approximate Arrhenian temperature dependence are defined as strong and specified with a lower value of F_i ($F_i \approx 16$), while the other side limit i.e. fragile glass forming liquids categorized by a higher value of F_i ($F_i \approx 200$). Thus, it is reasonable to state that the glasses having values of F_i within the above mentioned limit has been obtained from the good glass forming liquids.

5. Conclusions

In this chapter, use of calorimetry is discussed to understand structural relaxation in chalcogenide glasses. Some of theoretical approaches are also mentioned here that are generally used to estimate glass transition temperature before calorimetric experiment. Further, the usage of experimental data obtained from DSC or DTA for derivation of different parameters and apparent activation energy of glass transition region and hence understanding relaxation kinetics in glass transition region is also discussed in detail.

Author details

Balbir Singh Patial
Department of Physics, Himachal Pradesh University, Shimla, Himachal Pradesh, India

*Address all correspondence to: bspatial@gmail.com; bspatial@hpuniv.ac.in

References

[1] Mott NF, Davis EA. Electronic Processes in Non-crystalline Materials. Oxford: Clarendon Press; 1979. p. 200

[2] Odelevskii VI. The calculation of the generalized conductivity of heterogeneous systems. Journal of Technical Physics (USSR). 1951;**21**:678-685

[3] Kauzmann W. The nature of the glassy state and the behavior of liquids at low temperatures. Chemical Reviews. 1948;**43**:219-256. DOI: 10.1021/cr60135a002

[4] Zallen R. The Physics of Amorphous Solids. Weinheim: WILEYNCH Verlag GmbH & Co. KGaA; 1983

[5] Patial BS, Thakur N, Tripathi SK. On the crystallization kinetics of In additive Se–Te chalcogenide glasses. Thermochimica Acta. 2011;**513**:1-8. DOI: 10.1016/j.tca.2010.09.009

[6] Patial BS, Thakur N, Tripathi SK. Crystallization study of Sn additive Se–Te chalcogenide alloys. Journal of Thermal Analysis and Calorimetry. 2011;**106**(3):845-852. DOI: 10.1007/s10973-011-1579-5

[7] Anjali Patial BS, Bhardwaj S, Awasthi M, Thakur N. On the crystallization kinetics of multicomponent nano-chalcogenide $Se_{79\text{-}x}Te_{15}In_6Pb_x$ (x = 0, 1, 2, 4, 6, 8 and 10) alloys. Nano Express. 2020;**1**(3):030021. DOI: 10.1088/2632-959X/abc8c7

[8] Sharma N, Bhardwaj S, Thakur N, Patial BS. Crystallization study of Pb additive Se-Te-Ge nanostructured alloys using non-isothermal differential scanning calorimetry. Nanofabriaction. 2022

[9] Patial BS, Thakur N, Tripathi SK. A non-isothermal crystallization study of chalcogenide $Se_{85}Te_{15}$ glass using differential scanning calorimetry. Physica Scripta. 2012;**85**:045603. DOI: 10.1088/0031-8949/85/04/045603

[10] Abdel-Rahim MA, Hafiz MM, Shamekh AM. A study of crystallization kinetics of some Ge–Se–In glasses. Physica B: Condensed Matter. 2005;**369**(1-4):143-154. DOI: 10.1016/j.physb.2005.08.007

[11] Lafi OA, Imran MMA, Abdullah MK. Glass transition activation energy, glass-forming ability and thermal stability of $Se_{90}In_{10\text{-}x}Sn_x$ (x=2, 4, 6 and 8) chalcogenide glasses. Physica B: Condensed Matter. 2007;**395**:69-75. DOI: 10.1016/j.physb.2007.02.026

[12] Zheng Q, Zhang Y, Montazerian M, Gulbiten O, Mauro JC, Zanotto ED, et al. Understanding glass through differential scanning calorimetry. Chemical Reviews. 2019;**119**(13):7848-7839. DOI: 10.1021/acs.chemrev.8b00510

[13] Tichy L, Ticha H. Covalent bond approach to the glass-transition temperature of chalcogenide glasses. Journal of Non-Crystalline Solids. 1995;**189**:141-146

[14] Lankhorst MHR. Modelling glass transition temperatures of chalcogenide glasses. Applied to phase-change optical recording materials. Journal of Non-Crystalline Solids. 2002;**297**:210-219. DOI: 10.1016/S0022-3093(01)01034-1

[15] Tanaka K. Glass transition of covalent glasses. Solid State Communications. 1985;**54**:867-869. DOI: 10.1016/0038-1098(85)91158-5

[16] Lasocka M. The effect of scanning rate on glass transition temperature of splat-cooled $Te_{85}Ge_{15}$. Materials Science

and Engineering. 1976;**23**:173-177. DOI: 10.1016/0025-5416(76)90189-0

[17] Calventus Y, Suriñach S, Baró MD. Thermal stability and crystallization kinetics study of some Se-Te-Ge glassy alloys. Materials Science and Engineering A. 1997;**226-228**:818-822. DOI: 10.1016/S0921-5093(96)10801-7

[18] Mehta N, Singh KS. Effect of Sb and Sn additives on the activation energies of glass transition and crystallization in binary $Se_{85}Te_{15}$ alloy. Phase. Transit. 2009;**82**:43-51. DOI: 10.1080/01411590802260084

[19] Kissinger HE. Variation of peak temperature with heating rate in differential thermal analysis. Journal of Research of the National Bureau of Standards. 1956;**57**:217. DOI: 10.6028/jres.057.026

[20] Kissinger HE. Reaction kinetics in differential thermal analysis. Analytical Chemistry. 1957;**29**:1702-1706. DOI: 10.1021/ac60131a045

[21] Moynihan CT, Easteal AJ, Wilder J, Tucker J. Dependence of the glass transition temperature on heating and cooling rate. The Journal of Physical Chemistry. 1974;**78**:2673-2677. DOI: 10.1021/j100619a008

[22] Larmagnac JP, Grenet J, Michon P. Glass transition temperature dependence on heating rate and on ageing for amorphous selenium films. Journal of Non-Crystalline Solids. 1981;**45**(2):157-168. DOI: 10.1016/0022-3093(81)90184-8

[23] Kasap SO, Juhasz C. Kinematical transformations in amorphous selenium alloys used in xerography. Journal of Materials Science. 1986;**21**:1329-1340. DOI: 10.1007/BF00553271

[24] Hrubý A. Evaluation of glass-forming tendency by means of DTA. Czechoslovak Journal of Physics. 1972;**22**:1187-1193. DOI: 10.1007/BF01690134

[25] Saffarini G, Saiter A, Garda MR, Saiter JM. Mean-coordination number dependence of the fragility in Ge-Se-In glass-forming liquids. Physica B: Condensed Matter. 2007;**389**:275-280. DOI: 10.1016/j.physb.2006.06.163

[26] Štrbac GR, Petrović JS, Štrbac DD, Čajko K, Lukić-Petrović SR. Glass transition kinetics and fragility index of chalcogenides from Ag-As-S-Se system. Journal of Thermal Analysis and Calorimetry. 2018;**134**:297-306. DOI: 10.1007/s10973-018-7151-9

[27] Vilgis TA. Strong and fragile glasses: A powerful classification and its consequences. Physical Review B. 1993;**47**:2882-2885. DOI: 10.1103/PhysRevB.47.2882

Chapter 6

Cone Calorimetry in Fire-Resistant Materials

Rakesh Kumar Soni, Meenu Teotia and Aakansha Sharma

Abstract

Polymeric materials are specifically designed by compounding with additives to achieve specific properties that make them suitable for a particular application. Flame retardant materials offer fire-resistant properties to the polymers. The fire behavior of polymeric materials can be investigated with the help of various analytical techniques such as Underwriters Laboratories test standard UL-94, LOI test, Thermal gravimetric analysis (TGA) and Cone Calorimetry. Among these tests, Cone Calorimetry is the most suitable test method for predicting the real-scale fire behavior of polymeric materials and is adopted by the International Organization for standardization (ISO 5660-1). It quantifies heat generation, smoke production, mass loss and helps in the selection of polymeric materials for desired applications. In this chapter, an attempt has been made to present an overview of the thermal decomposition of polymers and the action of flame retardants. Different fire testing techniques generally used for investigations of fire characteristics of polymers are summarized.

Keywords: cone calorimetry, fire retardants, polymers, TGA, LOI

1. Introduction

Fire is a unique destructive force that thermally oxidizes polymeric materials. Fire-resistant materials can withstand high temperatures and resist burning. These may be inherently flame retardant or can be made fire resistant by adding flame retardants. Flame retardants are chemical substances that are incorporated in polymeric systems to combat fire risks hence the use of flame retardants in polymeric systems play a crucial role. Employment of flame retardants never signifies the absence of fire but flame retardants retard the ignition, growth and propagation of fire, thus minimizing fire-induced destruction [1, 2]. An increase in demands for polymers due to their wide applications in thermal, electrical and mechanical sectors has rekindled the flammability issues of flame retardants with polymers and created challenges for modern technology [3–5]. An exemplary flame-retardant system should have resistance to fire ignition, high combustion speed, smoke production and toxicity without economic penalty [2]. As polymers are organic compounds, they produce volatile combustible products when subjected to heat. Flame retardants interfere with the combustion process of polymers at various stages such as decomposition, ignition, the spread of flame and also smoke process [6]. Polymer flammability is defined by many processes and/or characteristics, including burning rates (solid degradation rate and heat release rate), spread rates (flame, pyrolysis, burn-out, smolder), ignition characteristics (delay time, ignition temperature, critical heat flux for ignition) [7], emission distribution (particularly toxic species emissions), smoke production. Several testing techniques are employed to analyze the fire behavior of polymers such as Underwriter's laboratories test standards (UL-94), limiting oxygen index (LOI) test, thermogravimetric analysis (TGA) and Cone calorimetry. In this chapter brief overview of the thermal decomposition of polymers and fire retardants & their mechanisms is presented under headings 2&3. Different types of flammability/fire tests used to analyze the fire characteristics of polymers are discussed under heading 4. It also covers the instrumental set up of the Cone calorimeter along with fire parameters obtained with the help of the instrument. Heading 5 concludes the advantages of Cone calorimetry over other analytical techniques used to determine fire characteristics of polymers.

2. Thermal decomposition of polymers

The physical and chemical properties of solid polymeric material change when heat is supplied to them as a result of thermal decomposition and thermal degradation. Thermal decomposition is a process in which extensive chemical changes can occur due to the application of heat. On the other hand, thermal degradation leads to loss of physical, mechanical and electrical properties of material when heat or elevated temperature is applied. To correlate the thermal properties of the material to the fire parameters, the process of thermal decomposition of the materials plays a crucial role. Thermal decomposition via chemical reactions generates gaseous flammable volatiles. When the flame is applied to polymeric substances, it tends to release flammable volatiles since polymer substances undergo pyrolysis. It releases some combustible gases, along with non-combustible gases (**Figure 1**). If the concentration of volatiles is at or above the combustible boundary then only it supports combustion otherwise flame dies out. Due to the presence of air, combustible products lead to flame and produce huge amounts of thermal energy, which in turn feed fire [8].

When thermal decomposition of deeper layers of solid polymeric material continues, the volatiles produced pass through the char layer formed to reach the surface. During this process, the char may cause secondary reactions to occur in the volatiles. Carbonaceous chars can be foamy layers, which slow down the decomposition of polymers considerably. Inorganic remains, on the other hand, can form glassy coverings that may then become impenetrable to volatiles and protect the beneath layers from any further thermal breakdown [9, 10]. Main chemical decomposition mechanisms are; (a) random-chain scission, in which chain scissions occur at random sites in the polymer backbone (b) end-chain scission, in which individual monomer units are consecutively removed at the chain end; (c) chain-stripping, in which atoms or groups not part of the polymer chain is cleaved; and cross-linking, in which bonds are created between polymer chains. Virgin polymer materials get decomposed by these processes. The reported literature indicates some type of correlation between mechanisms of polymer degradation and fire retardancy pathways. **Table 1** represents the polymer degradation mechanism of some synthetic polymers along with products evolved such as water, hydrocarbons and other volatiles.

The chemical processes involved in thermal decomposition are variably complex and the structure of polymers also influences the end consequence of decomposition. The chemical structure of repeating units and their heat-releasing capacity effects the fate of ignited polymer. Higher aliphatic molecules have a high potential for heat release which directly contributes to the thermal stability of the polymer [11].

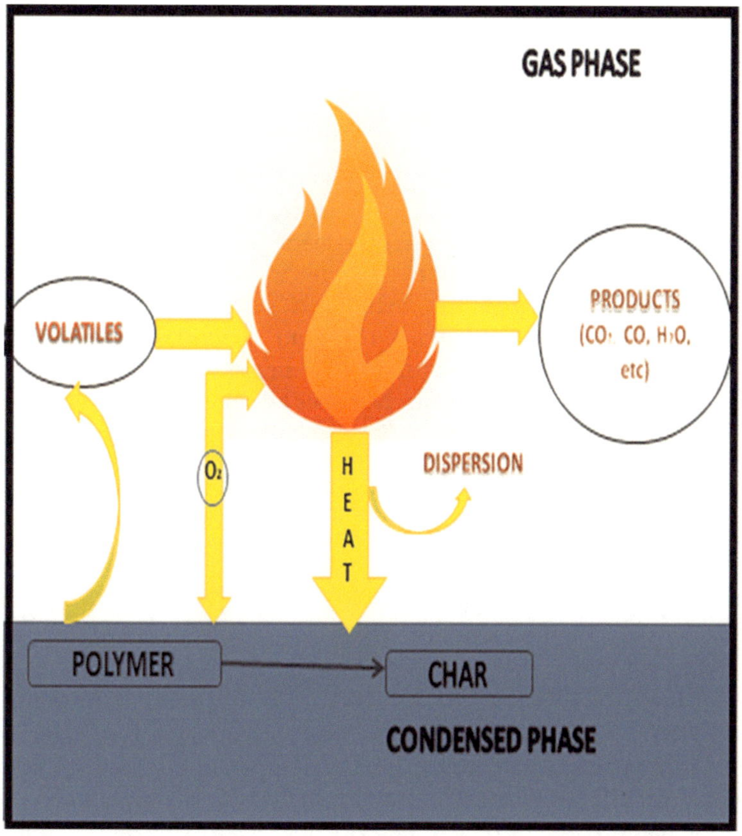

Figure 1.
Thermal decomposition of polymers.

Mechanism of polymer degradation	Examples of polymers	Products evolved	Flame retardant mechanism
Random chain scission	Polyethylene	Hydrocarbons	Vapor phase halogenated/ intumescent based
	Polypropylene	Hydrocarbons	Vapor phase halogenated/ intumescent based
	Polystyrene	Hydrocarbons	Condensed phase char formation
End chain scission	PMMA	Monomers	Halogenated FR (Free radical based)
	PFE	Monomers	Halogenated FR (Free radical based)
Chain stripping	PVC	Char, HCl, Hydrocarbons	Vapor phase brominated/ char formation
	PVA	Water, Char	Protective covering on material
Cross-linking	Polyacrylonitrile	Char and evolution of volatiles	Vapor phase / char formation
	Poly (oxy-m–xylene)	Char and evolution of volatiles	Vapor phase / char formation

Table 1.
Volatile products and flame retardant mechanism of polymers [12].

3. Fire retardants and their actions

Fire exists as a unique destructive force and results in damage to life and property if not handled carefully. Nowadays polymers are used in every walk of life and therefore their properties whether chemical or mechanical are very important parameters for their application in a particular product. In applications where fire retardance is an important requisite, polymers used must possess fire retardant ability to combat the unwanted risk of fire. Generally, polymers are made fire resistant by incorporating compounds, which are known as Flame retardants imparting flame retardance to the polymers. These can be thermoplastics, thermosets, fabrics, coating and prevent or inhibit the spread of fire. The word "flame retardant" describes a function rather than a chemical family. Flame retardants are made up of a range of compounds with various characteristics and structures, and these chemicals are frequently combined for increased efficacy [13, 14]. The action of flame retardants can be classified in three ways:

3.1 Vapor phase inhibition

When flame retardant chemicals react with the burning polymer during the radical gas phase, vapor phase inhibition occurs. These additives reduce or suppress the supply of combustible gases by interfering with the creation of free radicals, cooling the system, and lowering or suppressing the generation of free radicals. Brominated flame retardants (BFRs) are commonly utilized because of their vapor phase blocking characteristics. Before the material achieves its ignition temperature, bromine releases active bromine atoms into the gas phase. These atoms inhibit the chemical process occurring well within the flame, extinguishing or delaying its spread. This allows families or employees more time to flee the building or suppress the fire in another way.

3.2 Solid phase char formation

During a fire, solid-phase char flame retardants encourage the development of char. These additives reacted with the burning polymer, forming a carbon - containing layer on the surface of the material. This layer acts as a protective barrier, preventing the discharge of flammable gases and shielding the underneath material from the heat of such flame.

3.3 Quench and cool system

To increase flame resistance, quench and cool systems depend on hydrated materials. When hydrated minerals are exposed to fire, they emit water molecules that cool the substrate and interfere with the burning process.

4. Types of flammability/fire tests

Flammability testing is an important component of assuring safe and dependable consumer products. Textiles and consumer products, aircraft and transportation, mattresses, and furniture materials are among industries that use flammability test procedures. Flammability testing methods determine how readily materials ignite, burn, and respond when they are ignited. These include the following fire tests [15]:

4.1 UL-94 test (underwriters laboratories test standard UL-94)

One of the most commonly used flammability tests for estimating comparative flammability for plastic materials is UL 94 (Underwriters Laboratories test standard UL 94). Under specified laboratory circumstances, it analyzes a sample capacity to extinguish a flame after ignition as well as its dripping behavior in respect to a small naked flame or radiant source of heat [15]. Depending on the specifications, the materials are arranged vertically or horizontally over a Bunsen burner. During a vertical flammability test, the length of time a material burns after the initiating flame is withdrawn, the amount of the specimen that burns and whether or not it drops flaming particles are all measured. Horizontal flammability tests, on the other hand, look to see if the material burns after the test flame has been withdrawn, and then determine the rate at which the specimen burns. This fire test is intended to provide as a preliminary indicator of a plastic's acceptability to be used as part of an equipment or appliance in terms of flammability. The UL 94 standard is used to determine the intensity and characteristics of combustion based on the standard samples. UL-94 test is broadly classified into UL-94 VB and UL-94 HB [16].

The UL94 vertical burning test is a popular fire testing method for industrial polymeric products and materials. A 1/2" × 5" specimen is used for this test, which is held vertically at one end. The open end of the specimen is exposed to a burner flame at two 10-second periods, separated by the time required for the flaring combustion to stop after the first application [17]. The specimens are divided into two groups of five each. For each specimen, the following data is kept: (1) After the initial burner flame is applied, the duration of igniting combustion is calculated, (2) After the second burner flame is applied, the duration of blazing combustion is measured, (3). After the second burner flame is applied, the duration of glowing combustion is measured, (4) If burning droplets ignite cotton is put beneath the specimen and (5). If the specimen burns up while being held in the clamp.

Horizontal UL-94 Test is typically regarded as the easiest test to pass. The sample is clamped horizontally on the U-shaped frame and ignited in the combustion chamber for 15 seconds with a specified tiny flame. The flame is lit at one end of the specimen to determine the time it takes for the flame to spread to a certain extent and the distance at which it burns. A 1/2″× 5″ specimen is used in the test, which is held horizontally at one end with markings at 1″ and 5″ from the open end. For 30 seconds or until the flaming front reaches the 1-inch mark, a flame is applied to an open-end "a mark if the combustion persists, the timer is set between the 1" and 5″ marks. If the combustion process comes to a halt before the 5th minute, "The damaged length between the two markers, as well as the duration of combustion, is all recorded [16].

4.2 LOI (limiting oxygen index) test

Charles P. Fennimore and Fred J. Martin proposed the Limiting Oxygen Index (LOI) for the first time in 1966 [18]. It's a term that has been used to describe the relative flammability of various materials. The LOI test is subject to an international standard that is standardized in France (NF T 51-071) and the United States (ASTM D 2863) (ISO 4589). The limiting oxygen index (LOI) is the lowest oxygen concentration that will allow a polymer to burn. It is given as a percentage. It's determined by flowing a combination of nitrogen and oxygen over a burning material and gradually decreasing the oxygen level until the critical level is achieved. Standard tests, such as the ISO 4589 and ASTM D2863, are used to establish LOI values for various polymers. The limiting oxygen index is still the most important sorting criteria for polymer ignitability (LOI). The minimal oxygen content in the air for polymeric ignition and burning is determined by this method. As air contains 21% oxygen, polymers with LOI less than 21% are considered flammable, whereas those with LOI greater than 21% will not ignite in normal air. In reality, materials having an LOI of at least 25% should be less ignitable. Although, LOI is useful for proving polymer ignitability, it only provides limited evidence of material characteristics when exposed to heat, flames, or both [19, 20].

The test sample is vertically placed in a glass chimney, and then an oxygen/nitrogen atmosphere is created by a flow from the chimney's base. The flame is sparked at the top portion of the test sample, and thus the oxygen content in the flow is reduced until the flame can no longer be produced. The specimens are exposed to one or more specified sets of laboratory test conditions in this test technique. This is not always possible to detect changes in the fire-test-response characteristics assessed by or from this test if alternative test circumstances are substituted or the end-use conditions are modified. As an outcome, the results are only valid for the conditions indicated in the test procedure for exposure to fire [21].

4.3 TGA (thermal gravimetric analysis)

Thermo gravimetric analysis, often known as thermal gravimetric analysis (TGA), is a type of thermal analysis that measures the mass of a sample over time as the temperature varies. Physical events like as phase transitions, absorption, adsorption, and desorption, along with chemical phenomena such as chemical adsorption, thermal breakdown, and solid–gas interactions, are all revealed by this measurement (e.g., oxidation or reduction) [22]. The fundamental premise of thermo gravimetric analysis (TGA) has been that a sample's mass change may be evaluated under predetermined conditions. Absorption, adsorption, desorption, vaporization, sublimation, breakdown, oxidation, and reduction are all examples of thermal phenomena that TGA is used to explain [23]. Applications of TGA include, determination of thermal stability of materials (describes the breakdown mechanism, as well as the fingerprint materials, are used for identification and quality assurance), oxidative stability of materials, the composition of multi-component systems, estimated lifetime of a product, decomposition kinetics of materials, the effect of reactive or corrosive atmospheres on materials and moisture & volatiles content of Materials. This technique assists in investigations of fire characteristics of flame retardant polymers. Ureyen and Kayank investigated the effect of zinc borate with phosphorous-based FR finishing agents using cone calorimetry and TGA analysis [24]. TGA and DTA plots of zinc borate provided mass loss data of the samples which were further supported by heat release rate curves obtained from Cone Calorimetry. TGA measures thermal degradation and it can be related to flame resistive data of the samples.

4.4 Cone Calorimetery

The study of the thermal decomposition of polymers is important in terms of their fire performance behavior. This issue has been highlighted earlier by Van Krevelen [25], for many polymers, the limiting oxygen index [26] (primary test of flammability) could be linearly related to char yield as measured by Thermogravimetric analysis under specified conditions. The authors computed the char yield of various polymers with the help of structural parameters and found general low flammability linearly dependent on minimum thermal decomposition. Later on, some cases were noted in which low thermal stability and low flammability were found simultaneously in the polymeric substances. This raises a big question mark on the reliability of thermal decomposition of polymers and its correlation with the LOI test for predicting fire performance of polymeric materials [27, 28]. Later on, Lyon and Walters [29] gave a pathway to the preliminary prediction of fire performance of the polymers from heat release data derived from thermoanalytical data. Heat releasing capacity and the rate of heat release can be considered as important parameters to predict the fire performance of the material. Cone calorimetric analysis is considered one of the reliable techniques to comment on fire performance [30–35].

The fire community pushed for dependable bench-scale instruments to assess material flammability depending on heat release rate in the late 1970s and early 1980s. Heat release rates were believed to be the most dependable and accurate indicator of a material's flammability. The Cone Calorimeter, developed by NIST's Fire Research Division (formerly known as the Center of Fire Research at the National Bureau of Standards), was launched in 1982 as the next-generation device for determining material flammability [36].

A cone calorimeter is a device that is used to examine the condensed phase fire behavior of tiny samples of diverse materials. In the area of fire safety engineering, it is frequently utilized [37]. It collects information on the sample's burning characteristics, such as heat release rate, ignition time, combustion products, mass loss and other factors [38]. Huggett's principle states that the gross heat of combustion of any organic substance is proportional to the quantity of oxygen required for burning. This concept is used to calculate the heat release rate [39]. The Cone is a fire-testing device that works on the concept that the amount of heat generated by a burning sample is proportional to the amount of oxygen used during combustion. The intensity of a fire, such as fire development rate, is directly proportional to the amount of heat a substance generates.

4.4.1 Instrumentation setup

As it offers a lot of information with tiny samples, the cone calorimeter is the most commonly used device for studying the fire behavior of materials. ASTM E1354 and ISO 5660 have been used to standardize the procedure. The fundamental idea is to detect the decreasing oxygen content in flammable gasses of a sample ($100 \times 100 \times 4\ mm^3$) exposed to a certain heat flux (10–100 kW/m^2) [40]. A cone calorimeter is made up of several components and gadgets. Several parameters such as temperature, gas flow, mass and concentration are measured, logged, set, and adjusted by these parts when used together. The sample is mounted on a metal sample holder that is mounted on the load cell. During the experiment, the load cell records the sample's weight. Sample holders can be divided into two groups. The sample holder's edges may be open or closed, depending on the sample. If the sample is intumescent (that is, it swells and develops a protective char layer), a wired grid can be used to maintain it in place as it swells. When testing horizontally and with an edged framed sample holder, the wired grid is used. Below the cone heater is a spark igniter, which is placed directly above the sample surface. When the sample is heated, this ignites the flammable vapors that are escaping. The igniter is switched off and shifted to the side after the entire sample area has burned. A water-cooled heat flux measurement instrument is positioned on the sample surface level before the experiment during calibration. The temperature of the cone heater is then adjusted until the desired heat flux is achieved at the specimen's surface. The cone is usually positioned horizontally, although it can also be put vertically.

The form of the heater inspired the name of this testing device. Over 3 m long resistive heating wire packed with magnesium oxide refractory is coiled into a conical form to make the heater. The flue gases out from the flaming sample are gathered in an extraction chamber above the heater. The flow rate of combustion products is controlled by a flue gas fan installed in the flue gas line. In the flue gas line, the gas sampling ring is located before the fan. The gas sampling in the circle is first passed by two filters to remove pollutants, then via a cold trap and a drying agent to eliminate any potential water before contacting the gas analyzers. A smoke measuring system is located between both the gas sampling ring as well as the fan. A laser photometric beam detects the quantity of smoke generated [41, 42].

4.4.2 Important fire parameters determined by cone calorimeter

4.4.2.1 Heat release rate and Total heat release

Cone Calorimeter measures heat release rate at different incident heat fluxes. This measurement is based upon the principle that a nearly constant amount of heat is released per unit mass of oxygen consumed [43]. The concept of heat release rate calorimetry is simply the measurement of the rate at which a known weight (volume or area) of a carefully prepared specimen releases heat when exposed to a prescribed and controlled heating environment. Oxygen concentration in the flue gases is used to calculate the heat release rate. Oxygen consumed during combustion has been shown proportional to the heat released from a fuel [44]. Heat release rates and total heat released for a given time period are obtained from the strip chart record of changes in the fuel consumption required to maintain a constant temperature in the exhaust gases from the particular instrument. The traces record the decrease in fuel flow that is equivalent to the fuel contribution of the sample. An instantaneous heat release rate, expressed in watts/cm^2 area of the sample, is obtained by multiplying the net scale deflection at a given time by the calibration constant. The total heat release for a particular time interval was derived by graphic integration of the area under the curve between the desired time limits. The convention was established that zero time was the moment the sample "saw" the radiant flux in the exposure chamber (i.e., time was measured from the moment the door was closed). The average heat release rate for a given time period was obtained by determining the area under the curve (i.e., total heat released) and dividing by the time period of interest. In practice, this averaging is done for each successive 1-minute interval under the curve.

The heat release rate is an important parameter in determining the fire characteristics of materials. Holdsworth et al., in the year 2014 investigated different metal oxalates blended with Polyamide 6, 6 using cone Calorimetry at a heat flux of 50 kw/m^2 for peak heat release rates and found lowest PHRR (peak heat release rate) for halogenated composites showing better flammability resistance [45]. The authors further added metal complexes in Polyamide 6, 6 and obtained data for various fire parameters such as ignition time, THR, HRR and PHRR [46]. These formulations were also investigated for LOI values which were not found in agreement with Cone calorimetric data. This disagreement in LOI values with cone calorimetric data was indicated towards cross-linking and increment in char formation. Hence LOI values cannot be considered as sufficient data to comment upon fire characteristics of the material. Ramgobin et al., used the Cone Calorimetry technique to observe metal salen complexes as a fire retardant for thermoplastic polyurethane (TPU) [47]. Salen complexes of nickel, manganese and copper were fabricated in TPU and the comparative curves of heat release rates gave an idea of the best formulation based on PHRR and THR values. The lowest is PHRR and HR, better is the flame resistance performance of the sample. The width of the peak implies the release of combustibles at a broad range of temperature. Therefore, from THR data, it was found that formulations of TPU incorporated with metal complexes have at least 15% lower HR values than neat TPU. Recently, copper salen complexes were incorporated in PVC samples to study their flame resistive effect. It was found that from the cone calorimetric data that the complexes were improving fire retardance of PVC sheets only at 1 phr concentration. A decrease of 15.41 MJ/m^2 in total heat released was observed in PBr0 and of 10.42 MJ/m^2 in PBr1 sheets for the control sample sheet [48] as represented pictorially in **Figure 2**.

4.4.2.2 Char

The properties of the char and the underlying sample determine the time at which the stresses in the char layer are relieved by cracking and spalling, with subsequent reduction in the protective effects of the char layer. The finer and more uniform the texture of the sample, the longer the char layer is expected to remain intact. Schaffer showed that the charring rates of samples generally decrease as the density increases [49]. McLean attributed this effect to the variation of thermal diffusivity with density

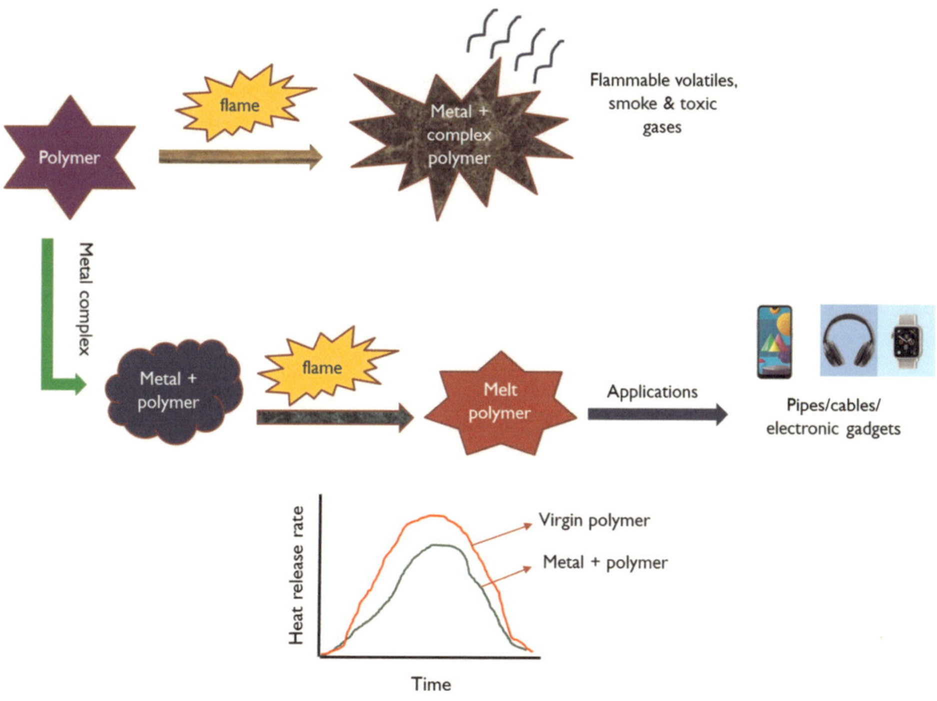

Figure 2.
Effect of fire retardants on the heat release rate.

[50]. Thermal diffusivity, a measure of how quickly a material absorbs heat from its surroundings, is defined as the ratio of thermal conductivity to the product of density and specific heat: a = (k/pC).

4.4.2.3 Mass loss rate

The rate of weight loss of a burning material has also been used as a measure of the rate of energy released during burning. The weight loss rate is directly related to the rate of heat release only in those cases where fuel composition and combustion efficiency remain constant as burning progresses [51].

4.4.2.4 Smoke release

Cone calorimeter also provides smoke production data which is often unused for predicting fire characteristics of polymers. Sonnier et al., in the year 2019 proposed new insights into the investigation of smoke production using a cone calorimeter [52]. The authors correlated smoke release to heat release using pure and flame retarded polymers, the carbon fractions and the presence of aromatic groups.

4.4.3 Interpretation of results by cone calorimeter

The cone calorimeter is an appropriate technique for evaluating the smoke suppression and fire retardance behavior of polymer samples. The cone calorimetric test is one of the most accurate methods for determining the fire resistance of polymeric materials. The overall heat release rate (HRR), peak heat release rate (PHRR), ignition duration, char yield, and smoke concentration of the sample are all determined by this test. The fire parameters can be determined by bench-scale test methods illustrated as arbitrary scale evaluation and FO- category models. The least correlation was found between these two test methods when applied on PVC panels and other polymeric materials.

Fire behavior analysis can be done through Petrella arbitrary scales, Östman and Tsantaridis model and Hansen & Hovde model [53, 54]. Two parameters, the flashover propensity ‘x’ (in kW/m^2s) and THR ‘y’ (in MJ/m^2) were proposed by Petrella for studying the contribution of the materials to flashover and thermal contribution [55]. The x and y values are helpful in determining the risk factor of the materials to thermal and flashover contribution under fire hazards.

$$x = \frac{peak(HGR)}{t_{ig}} \tag{1}$$

$$y = THR = \int_0^\infty HGR(t)dt \tag{2}$$

The total heat released (gross heating value, in MJ/m^2) is calculated by Eq. (3) and average heat generation rates; T60 (over 60 s After ignition), T180 (over 180 s After ignition) and T300 ((over 300 s After ignition (kW/m^2)) are calculated using Eqs. (4)–(6) respectively.

$$THR = \int_0^\infty HGR(t)dt \tag{3}$$

$$\overline{T}_{60} = \frac{1}{60}\int_{tig}^{tig+60} HGR(t)dt \tag{4}$$

$$\overline{T}_{180} = \frac{1}{180}\int_{tig}^{tig+180} HGR(t)dt \tag{5}$$

$$\overline{T}_{300} = \frac{1}{300}\int_{tig}^{tig+300} HGR(t)dt \tag{6}$$

Östman and Tsantaridis presented a relatively simple empirical linear regression model for the prediction of time to flashover in the room corner test. The model is based on empirical data, and was found to predict time to flashover with good accuracy for several products. Cone calorimeter results from tests at incident radiation heat flux of 50 kW/m^2 are used as input data to this model, which also requires information about the mean density of the tested product.

Cone calorimeter data can be used to derive useful information on studying the fire behavior of polymers. Based on the data and rational developed models, prediction of flashover time leading to fire behavior classification is feasible. Pyrolysis dynamics will help the understanding of the fundamental thermal behavior of materials leading to macroscopic fire behavior [56].

5. Advantages of cone calorimeter

Cone calorimetric test is one of the reliable tests to comment on the fire performance of polymeric substances. This test gives precise results about total heat release rate, peak heat release rate, ignition time, char yield, the smoke concentration of the sample. The best parameter for predicting fire hazard of a polymer is HRR (Heat Release Rate) in flaming combustion, although determination of HRR is a complex task for solid samples as it depends upon the heat flux, sample thickness, sample position, ventilation, etc. The justification for the widespread use of LOI test in its present form in flammability research is questionable. Aside from convenience and precision, it has many unfavorable features. Since it is a downward-burning test, this is a fire configuration of minor importance in real fires. Moreover, in the LOI test, characteristics of the heat transfer and rate of burning are fundamentally different from those in the much more important upward burning configuration. The LO1 has the unreal feature of being run at oxygen concentrations usually above the normal oxygen content of air, conditions which are probably important only in some space vehicles and oxygen contents in hospitals. It has poor thermal coupling between flame and substrate, and it is prone to severe disturbance by melt-flow phenomena. These are the ignition conditions well defined in the ASTM D2863 version of the oxygen index test. Probably the most severe deficiency of the LOI, is the lack of correlation with heat release results, is that it does not, except in unusual cases, predict the real fire performance of materials. It is still not clear whether any of the fire tests would provide enough energy feedback to stronger fire, to be able to predict full-scale fire performance, but they might provide better guidance for research and development than the LOI [57].

The development of a fire in the flame spread and sustained burning stages is dependent upon the rate of heat release and is not inherently related to the total heat that would be released if the specimen burned to completion. An important aspect of the flame spread and sustained burning stages is the production of smoke and toxic gases. These stages, culminating in flashover when the fire is in an enclosure, are dominant as far as life safety is concerned. The important aspect of the third stage is fire endurance, in which efforts to contain the fire and save the structure depend in part upon the rate of combustion or rate of heat release of building contents and materials of construction. Test methods to evaluate the performance of building materials for flame spread, smoke production, and fire endurance have assumed major importance in efforts to control fires. The rate of heat release is a relatively new criterion for evaluating fire behavior and is expected to assume similar importance in research, material assessment, and building regulations.

Cone Calorimetry is a reliable technique that involves the determination of one specific physical, chemical, or behavioral characteristic of a material, product, or system. A system test would characterize the overall behavioral reaction of a material, product, or system with the environmental as well as internal variables which influence its performance. A system test involves interactions between the material, product, or system with its surroundings. Heat release rate test methods are among those classified as system tests. Cone Calorimeter predicts real-scale fire behavior of materials. The obtained results are used in the ranking of products for fire performance and assist in the development of new fire-resistant materials and products [58]. The physical observations that can be visualized during the tests are surface rising, deformation owing to intumescences, residual stresses, the collapse of structures, char cracking, char development and cracking through bubbling and sparking, creation of solid crusts, tiny explosions, surface layers or bubbles, afterglow, and so on. Cone Calorimeter has been used widely for a variety of samples including ornamental plants [59], wood [40], textiles [28] and polyvinyl chloride products [48, 60]. Moisture content, physical properties and chemical composition are all variables that impact the flammability of live plants and can be analyzed with the help of Cone Calorimeter.

6. Conclusions

Polymeric materials are normally compounded with various compounds to enhance their properties such as tensile strength, flexibility, stability and fire resistance to be used in specified applications. These properties are analyzed using efficient analytical techniques to rank the polymeric materials. Fire resistance property is analyzed with the help of Underwriters Laboratories Test standard UL-94, LOI test, Thermal Gravimetric analysis and Cone calorimetry. Underwriters Laboratories Test is specially designed for the evaluation of those plastic materials that used in appliances, in response to a small, open flame or radiant heat source under controlled laboratory conditions and it cannot be used for polymeric materials used in construction, wall and floor coverings or other decorative objects. LOI test results are only valid for specified conditions indicated in the test procedure and cannot be applied to real scale conditions. Thermogravimetric analysis predicts the volatile components evolved during the decomposition of the polymeric sample along with decomposition temperature; however, the fire characteristics of the polymer can be studied with the help of Cone calorimetric analysis. Cone calorimetry is a key tool for the real scale analysis of fire-resistant polymers. The flame profile data suggests the fire hazards and is important for development of new fire-resistant materials.

Conflict of interest

The authors declare no conflict of interest.

Author details

Rakesh Kumar Soni*, Meenu Teotia and Aakansha Sharma
Department of Chemistry, Chaudhary Charan Singh University, Meerut, India

*Address all correspondence to: rksoni_rks@yahoo.com

References

[1] Bourbigot S, Le BM, Troitzsch J. Introduction. In: Troitzsch J, editor. Flammability Handbook. Munich: Hanser Verlag Publication; 2003. pp. 3-7

[2] Beard A, Angeler D. Flame retardants: Chemistry, applications, and environmental impacts. In: Handbook of Combustion. Weinheim: Wiley-VCH Verlag GmbH & Co. KGaA; 2010. DOI: 10.1002/9783527628148.hoc017. ISBN: 978-3-527-33449-1

[3] Lu SY, Hamerton I. Recent developments in the chemistry of halogen-free flame retardant polymers. Progress in Polymer Science. 2002;**27**(8): 1661-1712. DOI: 10.1016/S0079-6700(02)00018-7

[4] Rahman F, Langford KH, Scrimshaw MD, Lester JN. Polybrominated diphenyl ether (PBDE) flame retardants. Sci Total Environ. 2001;**275**(1-3):1-17. DOI: 10.1016/S0048-9697(01)00852-X

[5] Marosi G, Márton A, Anna P, Bertalan G, Marosföi B, Szép A. Ceramic precursor in flame retardant systems. In: Polymer Degradation and Stability. Vol. 77. Hungary: Elsevier; 2002. pp. 259-265. DOI: 10.1016/S0141-3910(02)00057-5

[6] Chen L, Wang YZ. A review on flame retardant technology in China. Part I: Development of flame retardants. Polymers for Advanced Technologies. 2010;**21**(1):1-26 https://onlinelibrary.wiley.com/doi/full/10.1002/pat.1550

[7] Bourbigot S, Duquesne S. Fire retardant polymers: Recent developments and opportunities. Journal of Materials Chemistry. 2007;**17**(22): 2283-2300 https://pubs.rsc.org/en/content/articlehtml/2007/jm/b702511d

[8] Kashiwagi T. Polymer combustion and flammability—Role of the condensed phase. Symposium on Combustion. 1994;**25**(1):1423-1437

[9] Staggs JEJ. Heat and mass transport in developing chars. Polymer Degradation and Stability. 2003;**82**(2):297-307

[10] Chen C, Yao B, Fan W, Liao G. A comparative experimental study on heat release rates of charring and noncharring solid combustible materials. Journal of Fire Sciences. 2016;**21**(5):369-382 https://journals.sagepub.com/doi/abs/10.1177/0734904103036091

[11] Boryniec S, Przygocki W. Polymer combustion processes. 3. Flame retardants for polymeric materials. Prog Rubber Plast Technol. 2001;**17**(2): 127-148 https://journals.sagepub.com/doi/abs/10.1177/147776060101700204

[12] Levchik SV, Weil ED. Overview of the recent literature on flame retardancy and smoke suppression in PVC. Polymers for Advanced Technologies. 2005;**16**(10):707-716 https://onlinelibrary.wiley.com/doi/full/10.1002/pat.645

[13] De Schryver D, Landry SD, Reed JS. Latest developments on the flame retardancy of engineering thermoplastics – SAYTEX® HP-7010 (brominated polystyrene) in glass filled engineering thermoplastics. Polymer Degradation and Stability. 1999;**64**(3): 471-477

[14] Iqbal A. Studying the Synergistic Effect of Aluminium Trihydroxide, Polyaniline and Other Fillers on EPDM Rubber. Lahore: Department of Chemical Engineering, COMSATS University; 2021 http://repository.cuilahore.edu.pk/xmlui/handle/123456789/2104

[15] Bosica R, Carosio F. Lignin-Based Nanocomposites with Fire-Retardancy Properties: Analysis and Implementation of an Innovative Material. KTH Royal Institute of Technology (SVEZIA): Politecnico di Torino; 2020. https://webthesis.biblio.polito.it/secure/cgi/set_lang?lang=it&referrer=https%3A%2F%2Fwebthesis.biblio.polito.it%2F15625%2F

[16] Yang C, Jílková SR, Melymuk L, Harris SA, Jantunen LM, Pertili J, et al. Are we exposed to halogenated flame retardants from both primary and secondary sources? Environmental Science & Technology Letters. 2020; 7(8):585-593. DOI: 10.1021/acs.estlett.0c00268

[17] Wang Y, Zhang F, Chen X, Jin Y, Zhang J. Burning and dripping behaviors of polymers under the UL94 vertical burning test conditions. Fire and Materials. 2010;**34**(4):203-215 https://onlinelibrary.wiley.com/doi/full/10.1002/fam.1021

[18] Koncar V. Smart Textiles for In Situ Monitoring of Composites. The Textile Institute Book Series Elsevier: Woodhead Publishing; 2018. pp. 1-151. ISBN: 9780081023099

[19] Bajaj P, Jha NK, Maurya PL, Misra AC. Flame retardation of polypropylene: Effect of organoantimony compounds on the structural and mechanical properties. Journal of Applied Polymer Science. 1987;**34**(5):1785-1801 https://onlinelibrary.wiley.com/doi/full/10.1002/app.1987.070340502

[20] Oxygen Index ASTM D2863. https://www.intertek.com/polymers/testlopedia/oxygen-index-astm-d2863/

[21] Sidebotham GW, Cross JA, Wolf GL. A test method for measuring the minimum oxygen concentration to support an intraluminal flame. ASTM Special Technical Publication. 1993;**1197**: 43-53 http://www.astm.org/DIGITAL_LIBRARY/STP/PAGES/STP24847S.htm

[22] Coats AW, Redfern JP. Thermogravimetric analysis. A review. The Analyst. 1963;**88**(1053):906-924 https://pubs.rsc.org/en/content/articlehtml/1963/an/an9638800906

[23] Sakho EHM, Allahyari E, Oluwafemi OS, Thomas S, Kalarikkal N. Dynamic light scattering (DLS). In: Thomas S, Thomas R, Zachariah AK, Kumar R, editors. Thermal and Rheological Measurement Techniques for Nanomaterials Characterization. Cambridge, United states: Elsevier; May 23, 2017. pp. 37-49

[24] Ureyen ME, Kaynak E. Effect of zinc borate on flammability of pet woven fabrics. Advances in Polymer Technology. 2019;**22**:1-13

[25] Van Krevelen DW. Some basic aspects of flame resistance of polymeric materials. Polymer. 1975;**16**(8):615-620

[26] Stauffer É, Lentini JJ. ASTM standards for fire debris analysis: A review. Forensic Science International. 2003;**132**(1):63-67

[27] Weil ED, Patel NG, Said MM, Hirschler MM, Shakir S. Oxygen index: Correlations to other fire tests. Fire and Materials. 1992;**16**(4):159-167 https://onlinelibrary.wiley.com/doi/full/10.1002/fam.810160402

[28] Schartel B, Hull TR. Development of fire-retarded materials—Interpretation of cone calorimeter data. Fire and Materials. 2007;**31**(5):327-354 https://onlinelibrary.wiley.com/doi/full/10.1002/fam.949

[29] Walters RN, Lyon RE. Molar group contributions to polymer flammability. Journal of Applied Polymer Science.

2003;**87**(3):548-563 https://onlinelibrary.wiley.com/doi/full/10.1002/app.11466

[30] Babrauskas V, Grayson SJ, editors. Heat Release in Fires. London,UK: Taylor & Francis; 1990

[31] Hirschler MM, Hoffmann DJ, Hoffmann JM, Kroll EC. Rate of heat release of plastic materials from car interiors. In: Proceedings of the 11th Annual Conference on Recent Advances in Flame Retardancy of Polymeric Materials. 2002. pp. 3-5 http://gbhint.tripod.com/papers_5_13_02/378_Cars_BCC2002.PDF

[32] Gann RG. Estimating data for incapacitation of people by fire smoke. Fire Technology. 2004;**40**(2):201-207 https://link.springer.com/article/10.1023/B:FIRE.0000016843.38848.37

[33] Fei Y, Jianjun Z, Yanghui Z, Peide L, Lin JZ, Chow TT. Preliminary real-scale experimental studies on cable fires in plenum. Journal of Fire Sciences. 2016; **21**(6):465-484

[34] Neviaser JL, Gann RG. Evaluation of toxic potency values for smoke from products and materials. Fire Technology. 2004;**40**(2):177-199 https://link.springer.com/article/10.1023/B:FIRE.0000016842.67144.12

[35] Peacock RD, Averill JD, Reneke PA, Jones WW. Characteristics of fire scenarios in which sublethal effects of smoke are important. Fire Technology. 2004;**40**(2):127-147 https://link.springer.com/article/10.1023/B:FIRE.0000016840.14947.61

[36] Zacharoff H. Simulating Cable Fires in Fire Dynamics Simulator: Based on Small Scale Testing in Cone Calorimeter. Bachelor Thesis X7007B 15 ECTS: Department of Civil, Environmental and Natural Resources Engineering Lulea University of Technology; 2021. http://urn.kb.se/resolve?urn=urn:nbn:se:ltu:diva-85995

[37] Twilley WH, Babrauskas V. User's Guide for the Cone Calorimeter. STIN: NASA STI/ Recon Technical Report N; 1988. Vol. 89 https://ui.adsabs.harvard.edu/abs/1988STIN...8922086T/abstract

[38] Nscort A. Estimation of Rate of Heat Release by Means of Oxygen Consumption Measurements. Intern Doc. Purdue University e-pubs: ALS-NCORT; 2003. https://docs.lib.purdue.edu/alsinternal/303

[39] https://en.wikipedia.org/wiki/Cone_calorimeter

[40] Yang X, Zhang W. Flame Retardancy of wood-polymeric composites. In: Polymer-Based Multifunctional Nanocomposites and Their Applications. 2019 Higher Education Press: Elsevier Inc; 2019. pp. 285-317

[41] Babrauskas V. Development of the cone calorimeter—A bench-scale heat release rate apparatus based on oxygen consumption. Fire and Materials. 1984;**8**(2):81-95 https://onlinelibrary.wiley.com/doi/full/10.1002/fam.810080206

[42] Lindholm J, Brink A, Hupa M. Cone Calorimeter—A Tool for Measuring Heat Release Rate. Turku, Finland: Åbo Akademi Process Chemistry Centre; 2009 http://ffrc.fi/FlameDays_2009/4B/LindholmPaper.pdf

[43] Janssens ML. Handbook of Environmental Degradation of Materials. 2nd ed. Elsevier, William Andrew; 2012

[44] Huggett C. Estimation of rate of heat release by means of oxygen consumption measurements. Fire and Materials. 1980; **4**(2):61-65. ISBN: 9781437734560

[45] Holdsworth AF, Horrocks AR, Kandola BK, Price D. The potential of metal oxalates as novel flame retardants and synergists for engineering polymers. Polymer degradation and stability. 2014; **110**:290-297

[46] Holdsworth AF, Horrocks AR, Kandola BK. Synthesis and thermal analytical screening of metal complexes as potential novel fire retardants in polyamide 6.6. Polymer Degradation and Stability. 2017;**144**:420-433

[47] Ramgobin A, Fontaine G, Penverne C, Bourbigot S. Thermal stability and fire properties of salen and metallosalens as fire retardants in thermoplastic polyurethane (TPU). Materials. 2017;**10**(6):665

[48] Teotia M, Verma A, Akitsu T, Tanaka S, Takahashi K, Soni RK. TGA decomposition and flame profile measurement of Terephthalamide stabilized PVC by cone calorimeter. Journal of Scientific and Industrial Research. 2017;**76**:438-441

[49] Schaffer EL. Review of Information Related to the Charring Rate of Wood. Vol. 145. United States Department of Agriculture, Research Note FPL-0145: Forest Products Laboratory; 1966

[50] JD ML. Thermal conductivity of wood. In: Heating, Piping and Air Conditioning. USA: Ashve Journal Section; Vol. 13. 1941. pp. 380-391

[51] https://nvlpubs.nist.gov/nistpubs/Legacy/IR/nbsir82-2597

[52] Sonnier R, Vahabi H, Chivas-Joly C. New insights into the investigation of smoke production using a cone calorimeter. Fire Technology. 2019; **55**(3):853-873

[53] Östman BA, Tsantaridis LD. Correlation between cone calorimeter data and time to flashover in the room fire test. Fire and Materials. 1994;**18**(4): 205-209 https://onlinelibrary.wiley.com/doi/full/10.1002/fam.810180403

[54] Hansen AS, Hovde PJ. Prediction of time to flashover in the ISO 9705 room corner test based on cone calorimeter test results. Fire and Materials. 2002; **26**(2):77-86 https://onlinelibrary.wiley.com/doi/full/10.1002/fam.788

[55] Petrella RV. The assessment of full-scale fire hazards from cone calorimeter data. Journal of Fire Sciences. 1994;**12**(1): 14-43 https://journals.sagepub.com/doi/abs/10.1177/073490419401200102?journalCode=jfse

[56] Xu Q, Jin C, Zachar M, Majlingova A. Test flammability of PVC Wall panel with cone calorimetry. Procedia Eng. 2013;**62**:754-759

[57] Schartel B, Braun U. Comprehensive fire behaviour assessment of polymeric materials based on cone calorimeter investigations. E-Polymers. 2003;**3**(1):1-14

[58] An W, Jiang L, Sun J, Liew KM. Correlation analysis of sample thickness, heat flux, and cone calorimetry test data of polystyrene foam. Journal of Thermal Analysis and Calorimetry. 2014;**119**(1): 229-238 https://link.springer.com/article/10.1007/s10973-014-4165-9

[59] Weise DR, White RH, Beall FC, Etlinger M. Use of the cone calorimeter to detect seasonal differences in selected combustion characteristics of ornamental vegetation*. Int J Wildl Fire. 2005;**14**(3):321-338 https://www.publish.csiro.au/wf/WF04035

[60] Soni RK, Sharma A, TAkitsu, T. and Teotia, M. Flame profile measurement of Cu (II) based Salen complex filled thermally stabilized PVC sheets by cone calorimeter. Journal of Scientific and Industrial Research. 2020;**79**(7):582-585

Section 3

Calorimetry in Biotechnology

Chapter 7

Calorimetry to Quantify Protein-Ligand Binding

Salerwe Mosebi

Abstract

Isothermal titration calorimetry (ITC) is the preferred method used to study biochemical reactions like protein-ligand binding due to its sensitivity, accuracy, and precision. ITC measures directly the heat absorbed or released (ΔH) associated with a given binding process. A typical ITC experiment allows the dissection of the binding energy of a reaction into ligand-enzyme association constant (Ka), change in enthalpy (ΔH), change in entropy (ΔS), change in Gibbs-free energy (ΔG), and the stoichiometry of association (N). The change in heat capacity (ΔCp) is obtained from the measurements of binding enthalpy over a range of temperatures. The magnitude and signs of the thermodynamic parameters that were obtained provide insight into the nature of interactions involved in the binding process. The strength of interaction is thermodynamically favorable is determined by the Gibbs free energy. ΔG is an important thermodynamic descriptor of a binding reaction since it dictates the binding affinity and is in turn defined by the enthalpy and entropy changes expressed in the following equation: $\Delta G = \Delta H–T\Delta S$. Up-close, this reflects the contradistinctions of two thermodynamic effects at a molecular level—the propensity to drop to lower energy (bond formation, negative ΔH), counterbalanced by the innate thermal Brownian motion's destructive characteristic (bond breakage, positive ΔS).

Keywords: isothermal titration calorimetry, binding energy, association constant, entropy change, enthalpy change, heat capacity

1. Introduction

The completion of the human genome project over 18 years ago has catapulted the number of novel targets for drug development to great heights. Many of these targets belong to protein families with homologous structures and similar binding pockets, which are crucial in regulating pathways and interaction networks describing cell function and inter-relation. It is also apparent that the basis of molecular recognition in drug discovery, signal-transduction, and protein-ligand complexes requires complete structural and thermodynamic dissection of macromolecular interactions involved. Several techniques (fluorescence, absorbance, nuclear magnetic resonance, surface plasmon resonance, biolayer interferometry, and ultracentrifugation) have been used as premier tools for characterizing interactions of biomolecules. These techniques can only determine the binding affinity constant (K_a) and indirectly derive other thermodynamic parameters. However, due to its sensitivity, accuracy,

IntechOpen

and precision, isothermal titration calorimetry (ITC) is the most rigorous and preferred method applied in a wide range of chemical and biochemical reactions. ITC has the advantage of directly quantifying the binding energetics of biological processes that include but not limited to protein–protein binding, protein-ligand binding, protein-DNA binding, protein-carbohydrate binding, protein-lipid binding, and antigen–antibody binding. ITC does this by measuring, directly and in real-time, the heat absorbed or released (ΔH) associated with a given binding process. A typical ITC experiment allows the dissection of binding energy of a reaction into ligand-enzyme association constant or binding affinity (K_a), change in enthalpy (ΔH), change in entropy (ΔS), change in Gibbs free energy (ΔG), the stoichiometry of association or number of binding sites (N), and the change in heat capacity (ΔC_p) obtained from measurements of binding enthalpy over a range of temperature. More importantly, ITC can be used to determine very low (10^3 M^{-1}) to very high association constants (10^{12} M^{-1}) without the need to use labels or immobilization of the binding components.

2. Fundamental principles of the ITC technique

A detailed description of the instrument and technique can be found elsewhere in the literature [1–5]. Briefly, the titration calorimeter consists of the injector system, adiabatic shield, and matched reference and sample cells (see **Figure 1a**). There is a self-stirring padded injection syringe and the thermostatic and feedback power systems that are computer-controlled. This instrument measures in real-time the thermal power that occurs when a solution in a syringe is titrated into a sample cell. In a typical ITC instrument, a pair of cylindrical cells (referred to as sample cell and reference cell) with volumes ranging from 200 to 1400 μl are present and contain analyte solution and reference buffer or water, respectively [6, 7]. The

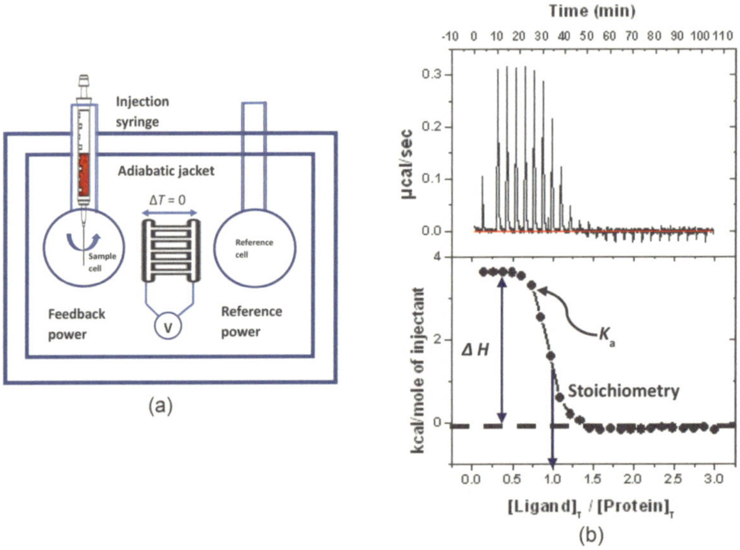

Figure 1.
*(**A**) A schematic representation of the ITC instrument setup, showing the sample and reference cells enclosed by a thermostated jacket. (**B**) an example of an ITC assay.*

thermostated adiabatic shield ensures that no heat exchange occurs between the cells and the surroundings [2]. The two cells are maintained at a constant and identical temperature by a feedback system that supplies thermal power continuously. In the event of a reaction in the sample cell usually accompanied by heat (exothermic reaction), the system ensures that the feedback power is withdrawn in order to retain thermal equilibrium between the cells. The feedback power supplied or withdrawn by electric resistive heaters located on the outer surfaces of the sample and reference cells to minimize temperature imbalances upon ligand injection is measured and converted into the heat of interaction. A sequence of injections is programmed and the ligand solution is injected at regular intervals into the sample cell through an automated injection syringe, which is stirred by rotation of the paddle-shaped syringe. After each injection (typically between 1 and 20 μl), the composition inside the sample cell changes causing the rearrangement of populations and complex formation [5]. Accordingly, as the series of injections continues, the system will experience various states of equilibrium each differing in composition. The heat released or absorbed with each injection corresponds to the increase in interacting species' concentration (as the reaction advances), and it is determined by the integration of the region under the deflection signal measured (amount of heat per unit of time provided to maintain thermal equilibrium in the sample and reference cells) [5]. If the binding between the injectant and the analyte is exothermic, this will result in the reduction in the power supplied by the feedback heater to maintain a constant temperature. On the other hand, if the binding is endothermic, there will be an increase in feedback power. At the end of the experiment, when no further heat is released or absorbed in the sample cell and saturation of the macromolecule is reached and it is possible to estimate K_a, ΔH, and N (independent variables). A typical result output of an ITC instrument is the feedback power measured as a function of time as shown in **Figure 1b**. The top panel represents the sequence of peaks as the solution in the syringe is injected into the analyte in the sample cell. The observed signal is the additional power that needs to be supplied or removed during the course of the experiment to keep a constant temperature in the sample cell and equal to the reference cell temperature. The reaction shown is that of an endothermic reaction, with an integrated heat plot in the bottom panel. Consequently, the areas under each peak, derived from per mole of ligand injected in each injection, are then plotted against the molar ratio of the total concentration of ligand to protein molecule concentration in the sample cell to obtain the following independent thermodynamic parameters: binding affinity, binding enthalpy, and the stoichiometry of binding. Notably, if two binding processes are characterized by different enthalpic and entropic terms and have the same Gibbs free energy of binding, they correspond to different binding modes, and therefore, the main underlying intermolecular interactions are different.

3. Protein-ligand binding energetics

As mentioned before, a typical ITC experiment allows the thermodynamic dissection of binding energy of a reaction into (K_a), (ΔH), (ΔS), (ΔG), (N), and (ΔC_p). Importantly, the magnitude and signs of the thermodynamic parameters obtained give us clues into the nature of interactions involved in the binding process, for example, the strength of the interaction and whether or not it is thermodynamically favorable is determined by the Gibbs free energy. If one ponders on the binding reaction under equilibrium conditions, where a macromolecule (P, protein) binds another molecule (L, ligand):

$$\mathrm{P}+\mathrm{L}\leftrightarrow PL \quad (1)$$

And if you assume that only one binding site is available, the association constant, K_a, which is inversely proportional to the dissociation constant, K_d, one is then able to determine the partition of the reactant molecules into free and bound species according to Eq. (2) below:

$$K_a=[PL]/[\mathrm{P}][\mathrm{L}] \quad (2)$$

The Gibbs free energy change of binding is an important thermodynamic descriptor of a binding reaction since it dictates the binding affinity or association constant:

$$\Delta G=-\mathrm{RT}\ln K_a \quad (3)$$

where R is the universal gas constant (8.314 J/mol/K), T is the temperature in kelvin, and K_a is the equilibrium binding constant. ΔG is in turn defined by the enthalpy and entropy changes and is expressed in the following equation:

$$\Delta G=\Delta H-\mathrm{T}\Delta S \quad (4)$$

At the molecular level, this reflects the contradistinctions of two thermodynamic effects at a molecular level—the propensity to drop to lower energy (bond formation, negative ΔH), counterbalanced by the innate thermal (Brownian) motion's destructive characteristic (bond breakage, positive $\Delta \mathrm{S}$) [8].

Since the native state of a protein exists as an ensemble of conformational states, the energy of stabilization of protein structure will not be evenly distributed throughout its three-dimensional structure [9]. There are regions of the protein with high stability constants (e.g., the hydrophobic core) and regions with low stability constants (e.g., loops and turns) with the majority of proteins exhibiting a dual character as originally observed for the HIV-1 protease [10, 11]. Since ligands with low molecular weight are in general not found attached to the exterior of proteins but are engulfed in crevices or binding pockets created by loops or other proteins' structural elements, the number of interactions between ligand and protein is increased and concomitantly enshrouds a substantial surface area from the solvent [9]. This conformational rearrangement often permits the entry of the ligand into the binding site and its subsequent shielding from the solvent; hence, makes favorable contributions to the Gibbs free energy of binding. If the rearrangements are only transient and the free and the bound states of the protein are similar, only binding kinetics are affected. If, however, the free and bound conformations of the protein are different, the binding affinity will be affected [9]. The Gibbs free energy associated with the change in protein conformation from its free to its bound state is included in the computation of the effective Gibbs energy of binding and corresponding binding affinity:

$$\Delta G_{\text{bind}} = \Delta G^{\circ}_{\text{bind}} + \Delta G_{\text{conf}} \tag{5}$$

where $\Delta G^{o}_{\text{bind}}$ is the Gibbs energy of binding obtained under the assumption that the free and bound states of the protein are the same, and ΔG_{conf} is the Gibbs energy associated with the change in protein conformation from its free to its bound form. In general, the Gibbs energy associated with a change from a less stable region to the bound conformation will be smaller than that associated with a change from a stable conformation to the bound conformation [9]. The presence of flexible regions in the protein molecule appears to facilitate the ligand-induced conformational changes if the putative binding site is not binding-competent in the ligand-free protein. The presence of regions with low stability also appears to provide a mechanism for achieving high binding affinity for low molecular weight ligands and serves as a starting point for the propagation of binding signals to distal sites [9].

Enthalpic and entropic contributions of the Gibbs energy originate from different types of interactions in the binding process. The binding enthalpy primarily reflects the energetic contribution of many individual interactions (hydrogen bonds, van der Waals interactions, polar, and dipolar interactions) between the ligand and the protein during the binding process, the conformational changes associated with binding, including interactions associated with the solvent. A negative (favorable) ΔH occurs when the interactions between the interacting molecules (e.g., hydrogen bond formation and van der Waals interactions) over-compensate the interactions of the individual molecules with the bulk solvent; otherwise, it will be positive (unfavorable, as for nonspecific hydrophobic interactions) [12]. The observed binding ΔH measured from a single ITC experiment often includes contributions not only from the actual binding event but also from the heat that is due to buffer ionization [13–15]. This is particularly true when the primary binding event is accompanied by the transfer of protons between the solvent and the protein-ligand complex. Thus, the determination of the intrinsic energetics of ligand binding requires experiments or measurements to be performed separately as a function of pH in buffers with different ionization enthalpies [13, 14, 16]. From this, pK_a values of ionizable groups responsible for proton linkage in the free and bound states and the number of protons that are coupled to the binding reaction can be easily calculated [14].

The binding entropy refers to the degree of disorder accompanying complex formation. Two major terms that contribute to the change in entropy are the solvation and conformational entropies. Solvation entropy arises from the gain in degrees of freedom of water molecules that, prior to the binding, are localized on the surface of the binding molecules and are released to the bulk solvent upon binding due to partial or complete desolvation of the two binding molecules. The change in solvation entropy is favorable (positive) if the surfaces that are buried upon binding are predominantly hydrophobic. It, therefore, originates from the burial of hydrophobic surfaces upon binding. Entropically driven ligand binding reactions are characterized by a large positive entropic contribution driven by the tendency of the molecule to escape water rather than by favorable interactions with the target molecule. In addition, the burial of solvent-exposed molecular surface area upon binding also contributes substantially to the heat capacity change upon complex formation due to the release of electro-restricted water or "hydrophobic water" from the binding site [17]. The conformational entropy, on the other hand, arises from changes in conformational degrees of freedom experienced by both the protein and the ligand upon binding. It is usually negative (unfavorable) due to the loss of degrees of freedom resulting from the reduction in the number of accessible conformations and configurations of both molecules (protein and ligand) upon binding.

4. Protein-ligand quantification and lead drug design

Currently, the development of lead compounds or drug design is centered on the optimization of their binding affinity toward the intended target. The binding affinity of a compound can be improved by generating a favorable binding enthalpy, favorable solvation entropy, and by minimizing the unfavorable conformational entropy. It is evident that simultaneous optimization of the three factors can achieve extremely high affinity. However, it is entirely feasible to design lead compounds that bind to the intended target with similar affinity but with different binding mechanisms, i.e., entropically or enthalpically driven ligands [18]. Entropically driven ligand derives most of its binding energy from a nonspecific hydrophobic effect, i.e., by making interactions of the drug with the solvent unfavorable, whilst enthalpically driven ligand derives its binding energy by establishing strong and specific hydrogen bonds with the target. Drug designers have long aimed at developing conformationally constrained ligands preshaped to the geometry of the selected binding site, which completes entropy optimization. Accordingly, a conformationally constrained molecule that is preshaped to the target achieves affinity, specificity, and selectivity through hydrophobicity and shape complementarity [19]. Perhaps, the most significant example is given by the development of the first-generation HIV-1 protease (HIV-1 PR) inhibitors (saquinavir, ritonavir, indinavir, and nelfinavir). The binding of these HIV-1 protease inhibitors is entropically driven and their binding enthalpy is either unfavorable (saquinavir, indinavir, and nelfinavir) or only slightly favorable (ritonavir) [20, 21]. In all cases, the dominant force for binding is a large positive entropy change that originates primarily from the burial of a large hydrophobic surface upon binding [20]. Moreover, since shape and hydrophobicity are nonspecific interactions, a change in the target binding site would lead to a reduction in the binding affinity. A low binding affinity reflects the inability of these conformationally rigid ligands to adapt to changes in the target binding pocket due to mutations or naturally occurring polymorphisms arising from genetic diversity. Hydrophobicity has historically been the preferred variable in the pharmaceutical industry due to its ease of implementation [22].

An enthalpically driven binding indicates specific interactions between two binding partners and corresponds well with ligand specificity, selectivity, and adaptability. Alternatively, an unfavorable enthalpic binding energy is indicative of nonspecific interactions between the binding partners, which in turn affects the ligand's specificity, selectivity, and adaptability. Despite apparent advantages of enthalpic interactions in achieving high affinity and improved selectivity, the optimization of the binding enthalpy has been more cumbersome to implement due to a large and unfavorable desolvation enthalpy of polar groups [23]. Generally, a polar group needs to make a strong interaction with the target in order to compensate for the desolvation enthalpy. Energetic contributions to binding affinity are not simply localized to the direct interactions between the molecules but contain interactions from structural and dynamic changes propagated throughout the protein, and from counter ions and hydrating water molecules located at the binding site. To be effective, an inhibitor needs to exhibit an extremely high affinity for the intended target and be mildly affected by mutations. Ideally, an inhibitor should have a binding affinity in the 1–50 pM range against the wild-type and be affected by mutations by a factor of 100 or less [24–26]. Compounds that achieve high binding affinity or that maximize binding affinity have been shown to combine or balance the favorable entropic and enthalpic contributions to the overall Gibbs energy of binding [27–30].

Notably, drug design paradigms have, to a large extent, illustrated how the enthalpic or entropic character of inhibitors is not dependent on the intended target, and that it is possible to develop entropically as well as enthalpically optimized inhibitors against the same binding site (e.g., HIV-1 protease). It has, for example, taken over 10 years to optimize HIV-1 protease inhibitors from the entropically driven inhibitors to the new and more potent enthalpically driven inhibitors [21, 24, 31]. The second-generation HIV-1 protease inhibitor, KNI-764 (AG-1776) for example, achieves the highest affinity (K_d = 32 pM) to the HIV-1 protease with a binding enthalpy (ΔH) of—7.6 kcal/mol and an entropic contribution ($-T\Delta S$) of −6.7 kcal/mol and can still afford the presence of certain flexible elements [21, 32]. The introduction of flexible asymmetrical functional groups in regions facing or in close proximity to mutation-prone areas of the protein provides adaptability to the inhibitor and low susceptibility to mutations [25, 26]. The increased conformational flexibility found in the second-generation HIV-1 protease inhibitors can also allow the inhibitor to compensate for the loss of interactions as a result of mutations in the target by burying a comparable or even larger surface area from the solvent [25].

New drug design strategies by calorimetric characterization have permitted the designers to recognize the nature of forces by which the HIV-1 proteins inhibitors bind the target primarily because these forces originate from different interactions. ITC was particularly crucial at the later stages since it gave a detailed description of the thermodynamic factors governing protein-inhibitor interactions essential for molecular recognition in HIV-1 protease binding and led to improvement in drug design. This task was also facilitated by structure-based algorithms able to predict the enthalpic and entropic consequences of introducing different functional groups in the lead compounds under investigation [9, 32]. Extensive studies using numerous techniques of molecular biology and the deepened understanding of drug-target at the molecular level have helped greatly in achieving rapid success in the area of drug development, especially in the treatment of AIDS [33–42].

5. Experimental approaches to determining the protein-ligand binding energetics using ITC

ITC experiments can be performed to determine the binding affinity, binding enthalpy, Gibbs free energy of binding, and stoichiometry of different inhibitors to the wild-type HIV-1 (South African subtype C (C-SA) protease. Indinavir, used in this study as an example, is an inhibitor that binds the wild-type HIV-1 protease with high affinity (with $Ki \leq 1$ nM). Therefore, the typical titration experiments are not able to accurately determine the association constant, even though the enthalpic contributions can be measured with high precision. A solution to this challenge is to perform calorimetric displacement titrations that will allow for the calculation of the binding affinity and enthalpy, as reported previously [25, 26, 43]. This technique allows complete determination of binding thermodynamics of high-affinity ligands ($Ka \geq 10^9$ M^{-1}) that are beyond the range of determination by direct titration. In calorimetric displacement titrations, the high-affinity inhibitor is titrated into a protease sample prebound to a weaker binding inhibitor (acetyl-pepstatin), a well-characterized inhibitor of lower binding affinity and unfavorable binding enthalpy [11]. The selection of a weak binding inhibitor with a binding isotherm of opposite sign (positive ΔH) produces a larger signal during the displacement reaction due to the displacement of the weaker binding inhibitor by an inhibitor yielding an exothermic isotherm (negative ΔH). As depicted in **Figure 2**, in the presence of the weak binding inhibitor, the apparent binding constant for the inhibitor which binds tightly, *K*app, falls within the range required for ITC determination. *K*app is given by Eq. (6) below:

$$Kapp = Ka / \left(1 + KB[B]\right) \tag{6}$$

where B is the concentration of the weaker binding inhibitor. In addition, *K*app can be lowered to the desired level by increasing the concentration of the weak

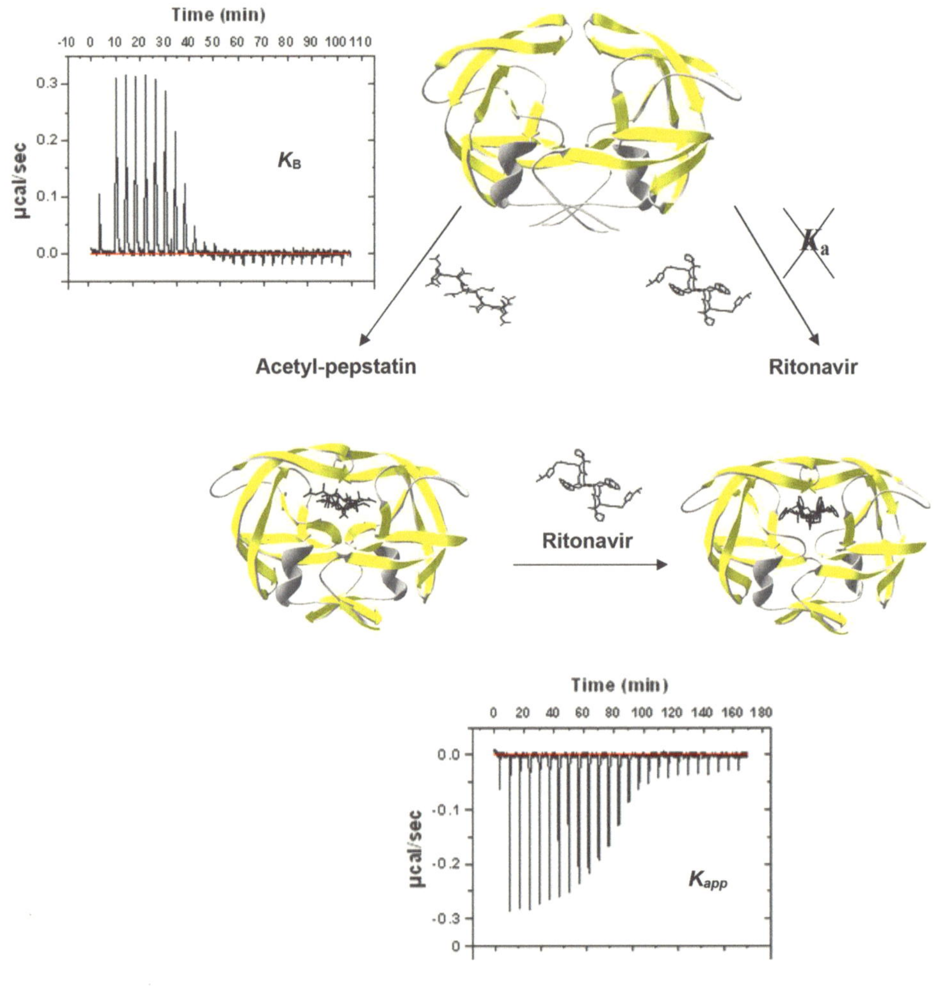

Figure 2.
Overview of a displacement titration assay for HIV-1 protease. The binding affinity of ritonavir, Ka, is beyond the limit of direct calorimetric determination. The displacement titration experiment is performed in the presence of the weak binding inhibitor acetyl-pepstatin (KB = 2.0 × 10^6 M^{-1}).

inhibitor. In addition, the binding isotherm, in this case, has sufficient curvature to allow for the calorimetric measurement of the apparent binding affinity of the stronger binding ligand [43]. Two calorimetric titrations need to be performed to work out the binding competition equations and calculate the association constant and enthalpic contributions of the tight-binding inhibitor: (1) titration of the weak binding inhibitor into the protease and (2) titration of the inhibitor of interest into the protease-(weak binding inhibitor) complex. The competition experiments were also performed at pH 5.0 using an acetate buffer with negligible binding enthalpy to minimize any proton coupling effect on the observed binding enthalpy [25].

Figure 3 shows typical displacement titrations for active site inhibitors of the wild-type C-SA HIV-1 protease in the presence of acetyl-pepstatin, pH 5.0. Each peak in the top panel represents the displacement of a weaker binding inhibitor (acetyl-pepstatin) from the active site of the protein by the tight-binding inhibitor (e.g., indinavir, with high binding affinity). As the titration progresses, the area under each peak becomes smaller due to increased occupancy of the available binding sites on the enzyme by the inhibitor of interest. The bottom panel in the figure shows the calorimetric binding isotherm obtained by plotting integrated heats obtained after each injection as a function of inhibitor concentration of interest per protein dimer. **Figure 3a** shows the integrated heats for the above peaks plotted against the molar ratio of acetyl-pepstatin to HIV-1 protease molecule. The solid line through the data represents the best fit using a one-site binding model. For the wild-type HIV-1 C-SA protease, the experimental data fit best to a

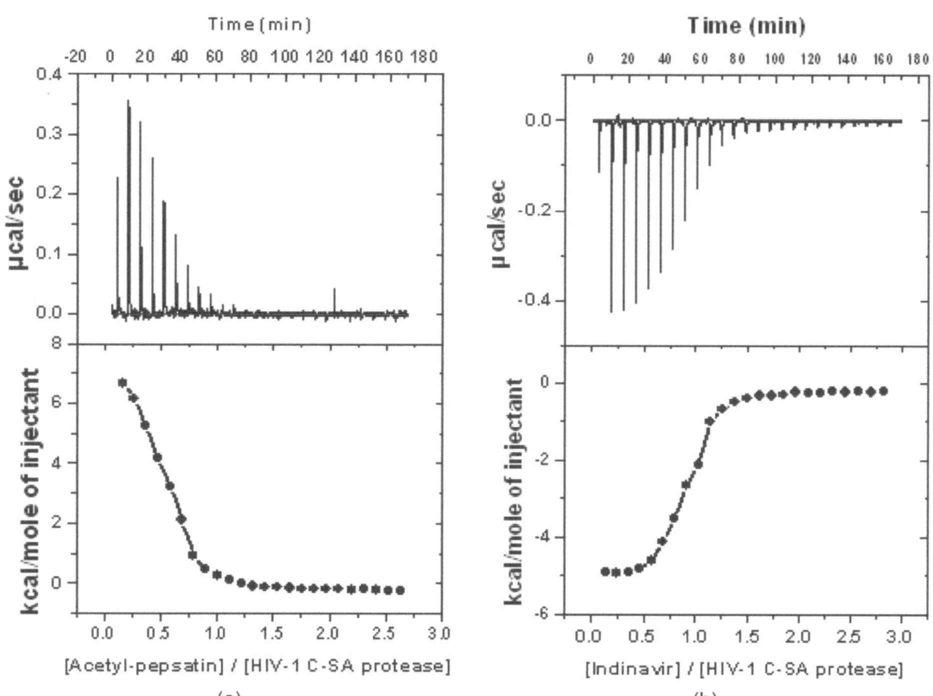

Figure 3.
(A) A representative calorimetric profile of the titration of the wild-type HIV-1 C-SA protease with acetyl-pepstatin. Titrations of acetyl-pepstatin (300 μM) into protease solution (20 μM). (B) ITC displacement calorimetric titration of indinavir (250 μM) into a solution of the wild-type HIV-1 C-SA protease (20 μM) prebound to acetyl-pepstatin (200 μM).

single-site displacement binding model; i.e., with the stoichiometry of 1:1 as shown in **Figure 3b**. The binding isotherms are monophasic with a sigmoidal fit to the data representing the decrease in available binding sites on the protein as the reaction progresses to completion. Used as a reference here, the clinical inhibitor, indinavir, binds to the wild-type C-SA HIV-1 protease with high binding affinity, *K*a, of $0.2 \times 10^9\ M^{-1}$ in a process strongly favored by entropic contributions, contributing about 12 kcal/mol to the overall Gibbs energy of binding at 25°C. At 25°C, the stoichiometry of binding for indinavir is 1.0 interpreted as one molecule of the inhibitor bound per dimer of HIV-1 protease and is consistent with crystallographic data [44–48]. Accordingly, the binding of indinavir is favored by entropic contributions of −15.0 kcal/mol, whereas its binding to the wild-type HIV-1 C-SA protease is characterized by a positive (unfavorable) enthalpy change of 2.70 kcal/mol. This is in agreement with the thermodynamic data obtained previously, which showed entropically controlled binding affinities and unfavorable or slightly favorable binding enthalpies [20, 49, 50]. Interestingly, for indinavir and other HIV-1 protease inhibitors like saquinavir, ritonavir, and nelfinavir, entropy ($-T\Delta S$) contributions as large as −16 kcal/mol, have also been shown by others to be required to compensate for the unfavorable binding enthalpies [20, 49].

6. Determination of the heat capacity change of a binding reaction

Although with a single ITC experiment, one is able to gain insights regarding the binding constant, binding enthalpy, binding entropy, stoichiometry of the reaction, and the Gibbs free energy of binding, another important parameter—the change in heat capacity (ΔC_p) upon binding can be obtained by performing the experiment at different temperatures and constant pressure. By applying Eq. (7) below, one can determine its value:

$$\Delta C_p = (\partial \Delta H / \partial T)_p \tag{7}$$

where $\partial\Delta H$ is the enthalpic change of binding at different temperatures (∂T) and ΔC_p is the change in heat capacity or slope obtained from plotting ΔH versus temperature. The heat capacity of a binding reaction is indicative of the burial of polar and nonpolar surfaces upon binding [51–53]. ΔC_p on an ITC instrument is typically obtained by measuring the enthalpic contributions of binding from 10–35°C at 5°C intervals without changing buffer and pH conditions. Although reports of the binding processes between a protein and a ligand have shown a negative and < 1 kcal/Kmol ΔC_p, the binding of two macromolecules (e.g., antigen–antibody) can induce higher heat capacity change, which is reflective of the burial of a larger solvent-accessible surface area as a result of the binding [26].

7. Conclusions

This chapter demonstrated the important role of calorimetry, in particular, isothermal titration calorimetry in dissecting the binding profile of two interacting species (e.g., a macromolecule and a ligand). It has obvious applications in drug development as it can be used for the characterization and optimization of lead compounds due to a wealth of thermodynamic information that is obtained from a single experiment. Some of the notable successes are in the lead optimization of HIV drugs exemplified by the HIV-1 protease discussed above. To this day, ITC remains a favored technique that can accurately characterize the interaction between the macromolecules and their biologically relevant binding partners. It is also uniquely positioned to assist us in getting a deepened thermodynamic understanding of the important biological processes in living systems like metabolism, active transport, biosensing, regulation, signal transduction, and integration to name a few.

Acknowledgements

The author would like to acknowledge the University of South Africa and the University of the Witwatersrand for the financial assistance and provision of resources and infrastructure needed to complete this work. The work was also supported by a grant from the National Research Foundation (Grant 121281 to S.M). Lastly, the author would like to thank Prof. Yasien Sayed from the University of the Witwatersrand for the invaluable supervisory role he played on the project.

Conflict of interest

The author declares no conflict of interest.

Author details

Salerwe Mosebi
University of South Africa, Florida, South Africa

*Address all correspondence to: mosebs@unisa.ac.za

References

[1] McKinnon IR, Fall L, Parody-Morreale A, Gill SJ. A twin titration microcalorimeter for the study of biochemical reactions. Analytical Biochemistry. 1984;**139**:134-139. DOI: 10.1016/0003-2697(84)90398-1

[2] Wiseman T, Williston S, Brandts JF, Lin LN. Rapid measurement of binding constants and heats of binding using a new titration calorimeter. Analytical Biochemistry. 1989;**179**:131-137. DOI: 10.1016/0003-2697(89)90213-3

[3] Freire E. Isothermal titration calorimetry: Controlling binding forces in lead optimization. Drug Discovery Today: Technologies. 2004;**3**:295-299. DOI: 10.1016/j.ddtec.2004.11.016

[4] Velazquez-Campoy A, Leavitt SA, Freire E. Characterization of protein-protein interactions by isothermal titration calorimetry. Methods in Molecular Biology. 2004;**261**:35-54. DOI: 10.1385/1-59259-762-9:035

[5] Velazquez-Campoy A, Freire E. ITC in the post-genomic era...? Priceless. Biophysical Chemistry. 2005;**115**:115-124. DOI: 10.1016/j.bpc.2004.12.015

[6] Malvern. Microcal ITC Systems: Understanding Biomolecular Interactions. Cambridge: Malvern; 2016

[7] TA. Microcalorimetry: ITC and DSC. Newcastle, DE: TA Instruments; 2019

[8] Cooper A. Thermodynamic analysis of biomolecular interactions. Current Opinion in Chemical Biology. 1999;**3**:557-563. DOI: 10.1016/s1367-5931(99)00008-3

[9] Luque I, Freire E. Structural stability of binding sites: Consequences for binding affinity and allosteric effects. Proteins. 2000;**4**:63-71. DOI: 10.1002/1097-0134(2000)41:4+<63::aid-prot60>3.3.co;2-y

[10] Luque I, Todd MJ, Gomez J, Semo N, Freire E. Molecular basis of resistance to HIV-1 protease inhibition: A plausible hypothesis. Biochemistry. 1998;**37**:5791-5797. DOI: 10.1021/bi9802521

[11] Todd MJ, Freire E. The effect of inhibitor binding on the structural stability and cooperativity of the HIV-1 protease. Proteins. 1999;**36**:147-156. DOI: 10.1002/(sici)1097-0134(19990801)36:2<147::aid-prot2>3.0.co;2-3

[12] Ross PD, Subramanian S. Thermodynamics of protein association reactions: Forces contributing to stability. Biochemistry. 1981;**20**:3096-3102. DOI: 10.1021/bi00514a017

[13] Gomez J, Freire E. Thermodynamic mapping of the inhibitor site of the aspartic protease endothiapepsin. Journal of Molecular Biology. 1995;**252**:337-350. DOI: 10.1006/jmbi.1995.0501

[14] Baker BM, Murphy KP. Evaluation of linked protonation effects in protein binding reactions using isothermal titration calorimetry. Biophysical Journal. 1996;**71**:2049-2055. DOI: 10.1016/S0006-3495(96)79403-1

[15] Baker BM, Murphy KP. Dissecting the energetics of a protein-protein interaction: The binding of ovomucoid third domain to elastase. Journal of Molecular Biology. 1997;**268**:557-569. DOI: 10.1006/jmbi.1997.097

[16] Doyle ML, Louie G, Dal Monte PR, Sokoloski TD. Tight binding affinities

determined from thermodynamic linkage to protons by titration calorimetry. Methods in Enzymology. 1995;**259**:183-194. DOI: 10.1016/0076-6879(95)59044-7

[17] Sturtevant JM. Heat capacity and entropy changes in processes involving proteins. Proceedings of the National Academy of Sciences of the United States of America. 1977;**74**:2236-2240. DOI: 10.1073/pnas.74.6.2236

[18] Velazquez-Campoy A, Todd MJ, Vega S, Freire E. Catalytic efficiency and vitality of HIV-1 proteases from African viral subtypes. Proceedings of the National Academy of Sciences of the United States of America. 2001;**98**:6062-6067. DOI: 10.1073/pnas.111152698

[19] Ruben AJ, Kiso Y, Freire E. Overcoming roadblocks in lead optimization: A thermodynamic perspective. Chemical Biology & Drug Design. 2006;**67**:2-4. DOI: 10.1111/j.1747-0285.2005.00314.x

[20] Todd MJ, Luque I, Velazquez-Campoy A, Freire E. Thermodynamic basis of resistance to HIV-1 protease inhibition: Calorimetric analysis of the V82F/I84V active site resistant mutant. Biochemistry. 2000;**39**:11876-11883. DOI: 10.1021/bi001013s

[21] Velazquez-Campoy A, Muzammil S, Ohtaka H, Schon A, Vega S, Freire E. Structural and thermodynamic basis of resistance to HIV-1 protease inhibition: Implications for inhibitor design. Current Drug Targets. Infectious Disorders. 2003;**3**:311-328. DOI: 10.2174/1568005033481051

[22] Lipinski CA. Drug-like properties and the causes of poor solubility and poor permeability. Journal of Pharmacological and Toxicological Methods. 2000;**44**:235-249. DOI: 10.1016/s1056-8719(00)00107-6

[23] Cabani S, Gianni P, Mollica V, Lepori L. Group distribution to the thermodynamic properties of non-ionic solutes in dilute aqueous solution. Journal of Solution Chemistry. 1981;**10**:563-595. DOI: 10.1007/BF00646936

[24] Velazquez-Campoy A, Freire E. Incorporating target heterogeneity in drug design. Journal of Cellular Biochemistry. Supplement. 2001;**37**:82-88. DOI: 10.1002/jcb.10068

[25] Velazquez-Campoy A, Kiso Y, Freire E. The binding energetics of first- and second-generation HIV-1 protease inhibitors: Implications for drug design. Archives of Biochemistry and Biophysics. 2001;**390**:169-175. DOI: 10.1006/abbi.2001.2333

[26] Ohtaka H, Schon A, Freire E. Overcoming drug resistance in HIV-1 chemotherapy: The binding thermodynamics of amprenavir and TMC-126 to wild-type and drug-resistant mutants of the HIV-1 protease. Protein Science. 2002;**11**:1908-1916. DOI: 10.1110/ps.0206402

[27] Ohtaka H, Velazquez-Campoy A, Xie D, Freire E. Multidrug resistance to HIV-1 protease inhibition requires cooperative coupling between distal mutations. Biochemistry. 2003;**42**: 13659-13666. DOI: 10.1021/bi0350405

[28] Ohtaka H, Muzammil S, Schon A, Velazquez-Campoy A, Vega S, Freire E. Thermodynamic rules for the design of high affinity HIV-1 protease inhibitors with adaptability to mutations and high selectivity towards unwanted targets. The International Journal of Biochemistry & Cell Biology. 2004;**36**:1787-1799. DOI: 10.1016/j.biocel.2004.02.021

[29] Ohtaka H, Freire E. Adaptive inhibitors of the HIV-1 protease. Progress in Biophysics and Molecular

Biology. 2005;**88**:193-208. DOI: 10.1016/j.pbiomolbio.2004.07.005

[30] Carbonell T, Freire E. Binding thermodynamics of statins of HMG-CoA reductase. Biochemistry. 2005;**44**:11741-11748. DOI: 10.1021/bi050905v

[31] Muzammil S, Armstrong AA, Kang LW, Jakalian A, Bonneau PR, Schmelmer V, et al. Unique thermodynamic response of tipranavir to human immunodeficiency virus type 1 protease drug resistance mutations. Journal of Virology. 2007;**81**:5144-5154. DOI: 10.1128/JVI.02706-06

[32] Vega S, Kang L, Velazquez-Campoy A, Kiso Y, Amzel LM, Freire E. A structural and thermodynamic escape mechanism from a drug resistant mutation of the HIV-1 protease. Proteins. 2004;**55**:594-602. DOI: 10.1002/prot.20069

[33] DeVita VTJr, Broder S, Fauci AS, Kovacs JA, Chabner BA. Developmental therapeutics and the acquired immunodeficiency syndrome. Annals of Internal Medicine. 1987;**106**:568-581. DOI: 10.7326/0003-4819-106-4-568

[34] Fitzgerald PM, Springer JP. Structure and function of retroviral proteases. Annual Review of Biophysics and Biophysical Chemistry. 1991;**20**:299-320. DOI: 10.1146/annurev.bb.20.060191.001503

[35] Kohlstaedt LA, Wang J, Friedman JM, Rice PA, Steitz TA. Crystal structure at 3.5 Å resolution of HIV-1 reverse transcriptase complexed with an inhibitor. Science. 1992;**256**:1783-1790. DOI: 10.1126/science.1377403

[36] Martin JA. Recent advances in the design of HIV proteinase inhibitors. Antiviral Research. 1992;**17**:265-278. DOI: 10.1016/0166-3542(92)90022-w

[37] Fitzgerald PMD. HIV protease-ligand complexes. Current Opinion in Structural Biology. 1993;**3**:868-874. DOI: 10.1016/0959-440X(93)90150-J

[38] Jacobo-Molina A, Ding J, Nanni RG, Clark AD Jr, Lu X, Tantillo C, et al. Crystal structure of human immunodeficiency virus type 1 reverse transcriptase complexed with double-stranded DNA at 3.0 Å resolution shows bent DNA. Proceedings of the National Academy of Sciences of the United States of America. 1993;**90**:6320-6324. DOI: 10.1073/pnas.90.13.6320

[39] Wlodawer A, Erickson JW. Structure-based inhibitors of HIV-1 protease. Annual Review of Biochemistry. 1993;**62**:543-585. DOI: 10.1146/annurev.bi.62.070193.00255

[40] Dyda F, Hickman AB, Jenkins TM, Engelman A, Craigie R, Davies DR. Crystal structure of the catalytic domain of HIV-1 integrase: Similarity to other polynucleotidyl transferases. Science. 1994;**266**:1981-1986. DOI: 10.1126/science.7801124

[41] Bujacz G, Alexandratos J, Zhou-Liu Q, Clement-Mella C, Wlowader A. The catalytic domain of human immunodeficiency virus integrase: Ordered active site in the F185H mutant. FEBS Letters. 1996;**398**:175-178. DOI: 10.1016/s0014-5793(96)01236-7

[42] Vacca JP, Condra JH. Clinically effective HIV-1 protease inhibitors. Drug Discovery Today. 1997;**2**:261-272. DOI: 10.1016/S1359-6446(97)01053-2

[43] Sigurskjold BW. Exact analysis of competition ligand binding by displacement isothermal titration calorimetry. Analytical Biochemistry. 2000;**277**:260-266. DOI: 10.1006/abio.1999.4402

[44] Baldwin ET, Bhat TN, Gulnik S, Liu B, Topol IA, Kiso Y, et al. Structure of HIV-1 protease with KNI-272, a tight-binding transition-state analog containing allophenylnorstatine. Structure. 1995;**3**:581-590. DOI: 10.1016/s0969-2126(01)00192-7

[45] Chen Z, Schock HB, Hall D, Chen E, Kuo LC. Three-dimensional structure of a mutant HIV-1 protease displaying cross-resistance to all protease inhibitors in clinical trials. The Journal of Biological Chemistry. 1995;**270**:21433-21436. DOI: 10.1074/jbc.270.37.21433

[46] Hong L, Treharne A, Hartsuck JA, Foundling S, Tang J. Crystal structures of complexes of a peptidic inhibitor with wild type and two mutant HIV-1 proteases. Biochemistry. 1996;**35**:10627-10633. DOI: 10.1021/bi960481s

[47] Hong L, Zhang XJ, Foundling S, Hartsuck JA, Tang J. Structure of a G48H mutant of HIV-1 protease explains how glycine-48 replacements produce mutants resistant to inhibitor drugs. FEBS Letters. 1997;**420**:11-16. DOI: 10.1016/s0014-5793(97)01477-4

[48] Hong L, Zhang XC, Hartsuck JA, Tang J. Crystal structure of an in vivo HIV-1 protease mutant in complex with saquinavir: Insights into mechanisms of drug resistance. Protein Science. 2000;**9**:1898-1904. DOI: 10.1110/ps.9.10.1898

[49] Mosebi S, Morris L, Dirr H, Sayed Y. Active site mutations in the south African HIV-1 subtype C protease have a significant impact on clinical inhibitor binding: A kinetic and thermodynamic study. Journal of Virology. 2008;**82**: 11476-11479. DOI: 10.1128/JVI.00726-08

[50] Maseko S, Padayachee E, Maphumulo S, Govender T, Sayed Y, Maguire G, et al. Kinetic and thermodynamic characterisation of HIV-protease inhibitors against E35D↑G↑S mutant in the south African HIV-1 subtype C protease. Journal of Enzyme Inhibition and Medicinal Chemistry. 2019;**34**:1451-1456. DOI: 10.1080/14756366.2019.1636234

[51] Privalov PL, Makhatadze GI. Contribution of hydration and non-covalent interactions to the heat capacity effect on protein unfolding. Journal of Molecular Biology. 1992;**224**:715-723. DOI: 10.1016/0022-2836(92)90555-x

[52] Murphy KP, Freire E. Thermodynamics of structural stability and cooperative folding behavior in proteins. Advances in Protein Chemistry. 1992;**43**:313-361. DOI: 10.1016/s0065-3233(08)60556-2

[53] Gomez J, Hilser VJ, Freire E. The heat capacity of proteins. Proteins. 1995;**22**:404-412. DOI: 10.1002/prot.340220410

Chapter 8

Calorimetry in Allergy Diagnostic

Evgeni Stanev and Maria Dencheva

Abstract

Calorimetry is an indisputable diagnostic method. Over the years, there has been an improvement in the equipment and methods for measuring the calor that accompanies various processes. Using a thermal camera, we can measure the surface temperature of the skin at the beginning and the end of each skin allergy test. They are epicutaneous, cutaneous, and percutaneous. In case of a positive reaction, allergic inflammation is observed with the obligatory symptoms, one of which is warming (calor). Measuring and visualizing this warming is essential in the diagnosis of allergic reaction. The methodology of imaging the skin areas and processing the results is the key point in the objectivity of the study. Diagnostic skin allergy tests report mainly immunopathological reactions of the first and fourth types (Coombs and Gel classification). Their course is different and this necessitated the development of various thermovisiographic imaging methods. Through the results of our thermal imaging studies, we derived a scale, that determines the intensity of the allergic reaction, for each of the skin allergy tests. The use of thermovisiography in addition to the standard reporting of allergic skin reactions provides precision and more information about the subtle temperature changes that accompany allergic reactions.

Keywords: allergic inflammation, thermocamera, local anesthetics, allergens, haptens, prick test, patch test, intradermal test

1. Introduction

Diagnosis in allergology is based on anamnestic data, clinical and laboratory tests. Despite the development of *in vitro* tests, skin allergy tests are used as the gold standard in clinical practice. They aim to provoke a local allergic reaction in the study area. Allergic inflammation is characterized by the same five signs of inflammation—tumor, rubor, calor, dolor, and functiolesa. In standard clinical practice, the strength of the reaction is determined by the size of the erythema (rubor) and the size of the papule (tumor). With the help of modern thermal imaging cameras, it is possible to take into account another parameter of the inflammatory reaction—calor. The warming covers the area of skin that has been in contact with the allergen or the hapten.

Skin allergy tests are easy to perform, inexpensive, quick to read, and ideal for diagnosis even in patients with limited mobility, but standard reporting carries with it a degree of subjectivity.

Methodology and implementation of the test:

The test is usually performed on the volar side of the forearm skin according to the skin prick test—European standards. The skin is cleansed with an alcohol swab. The places of application of the allergen—most often inhaled household, pollen allergens, insect and food allergens, some drugs, local and general anesthetics, iodine contrast agents, and vaccines—are marked with a skin marker. Against each marking is placed one drop of the substance. With a separate plastic lancet, the skin is pricked. Positive and negative controls are obligatory in this test. Conventional reading is after 20 minutes [1].

Method of performing an intradermal test:

In the intradermal test, a certain amount of the allergen is injected into the deeper skin layers. The reading is also 20 minutes [1].

The results of controlled skin sensitization in Prick and the intradermal test are reported by measuring in millimeters the diameters of the papule and erythema that form.

Methods of epicutaneous testing:

Testing with hypoallergenic patches, which are most often placed on the back. This is a test method for proving contact allergy/contact dermatitis, metal allergy, drug allergy, dental materials. The set of allergens for epicutaneous testing is applied separately in special chambers on the test patches, then glued to the back for a period of 48 hours.

It is reported to be a cell-mediated mechanism (Coombs and Gell type 4) of reaction on the 3rd and sometimes on the 7th day [2, 3].

The clinician needs to have accumulated practical knowledge and experience to be able to correctly report skin reactions. With the inclusion of thermal imaging analysis, this process is supported, and the image from the thermal imager can be objectively analyzed. The method is rapid, non-invasive, and may accompany standard reporting of allergic skin reactions [4].

The thermal imaging image provides information about the size of the inflamed area and its temperature. There are different types of thermal cameras, but FLIR T620 can be used for clinical practice. The analysis is performed using specialized software, as well as comparisons between the reaction to the allergen and reactions to the positive control, negative control, or intact skin area.

After performing the test and standard reading, a thermal camera is taken of the skin areas of interest. The thermal imaging methodology developed by us is different for each of the three skin allergy tests. The image is analyzed by extracting certain (basic) temperature parameters for each zone. Based on them, the additional parameters are calculated, which determine whether the reaction is positive and what is its intensity, which in turn supports the standard reading and minimizes the possibility of misinterpretation of the result.

In this chapter, we will present each of the three allergy tests in turn. Approaches to temperature analysis will be proposed for each test, which will allow quantitative measurement of the calor due to allergic inflammation. The conclusion will summarize the limit values of the indicators and their clinical significance in the reporting of tests.

2. Thermal imaging analysis of allergic skin reactions in prick test

In the Prick test, two thermal imaging images are taken—one before it is performed and one after it. It is necessary to observe the conditions for thermovisiographic imaging [5]. In standard reading, clinicians touch the skin to sense the presence and size of the papule. It is necessary to do this after the second scan, so as not to affect the temperature of the examined skin area. Temperature analysis requires measuring the change in each area of the skin. To achieve this, the hand needs to be positioned in the same way in both shots. The stand created by us (**Figure 1**) allows the patient's arm to be placed comfortably, and the curved shape of the two beds for the forearm to stabilize it in a certain position, regardless of its anatomical features. In addition to support elements, the stand includes a plastic template with five holes. The pattern fixation system above the volar surface of the forearm allows the arm to be removed and then repositioned, leaving the pattern position unchanged. The examined five skin areas can be marked with a skin marker after applying the template. The distance between the edges of each hole is 10 mm, enough to prevent the mixing of skin reactions between two adjacent tests (**Figure 2**).

The strongest skin reaction observed after the Prick test is to the positive control. Its size varies between 10 and 30 mm [6]. The diameter of the holes in the template is 20 mm, which may not, in some cases, cover the full size of the reaction to a histamine solution. However, this size is large enough for the temperature reactions of the negative control and the tested allergens.

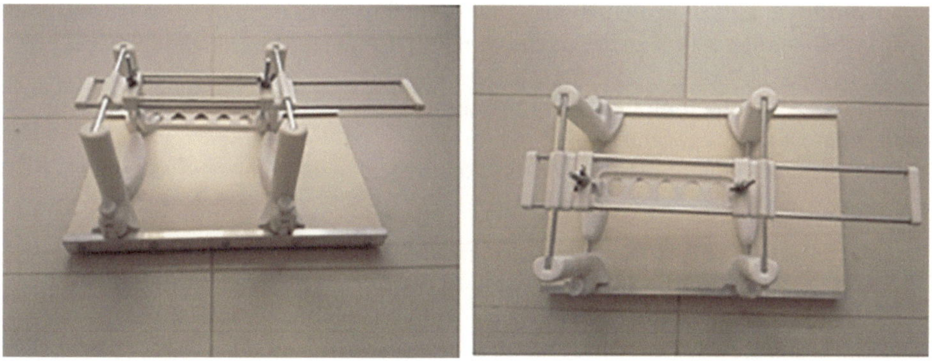

Figure 1.
Stand for fixing the patient's forearm during the prick test.

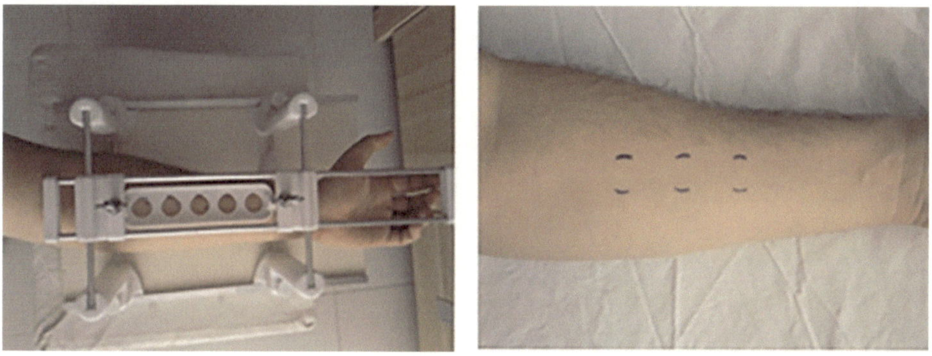

Figure 2.
Placing the patient's hand and outlining three skin areas (depending on the number of tested allergens) in which the prick test will be performed.

2.1 Analysis of the temperature of the skin areas BEFORE performing a prick test

Before conducting the test, the places where it will be performed are determined. The initial temperature of the skin areas on which the reactions will be observed should be uniform, without large amplitudes. These are observed in cases of superficially located large blood vessels (v. Cephalica; v. Basilica; v. Mediana anterbrachii). For this reason, it is necessary to initially capture the skin areas and reposition the arm (**Figure 3**). Most studies do not analyze the initial skin temperature, although it is the starting point for the temperature change [6–8]. There were no statistically significant differences in skin temperatures prior to the test when subcutaneous main blood vessels were avoided [9]. If the blood vessel passes through the examined area, it will not allow significant changes in temperature to be reported, regardless of the severity of the allergic inflammation. On the other hand, there is a risk of injury to the blood vessel and compromise of the allergy test [10].

The Prick test is most commonly used to test for sensitization to local anesthetics, foods, and medications. Before the test, two or three anesthetics are selected to be checked for sensitization.

The initial survey gives us information about the following main temperature parameters (**Figure 4**). With the help of specialized software, the skin areas of interest are outlined, and then a table with the temperatures of each of them is displayed.

X_1 – the average temperature of the skin area in which the test allergen will be located before the test.

Neg_1 – the average temperature of the skin area in which the negative control before the test will be located

Pos_1 – the average temperature of the skin area in which the positive control will be located before the test

2.2 Analysis of the temperature of the skin areas AFTER performing a prick test

After shooting, the test is performed in the standard way [10]. Before proceeding to the standard reading, the hand is carefully placed on the stand and the template is positioned so that the openings coincide with the marked skin areas. The second thermal imaging is done.

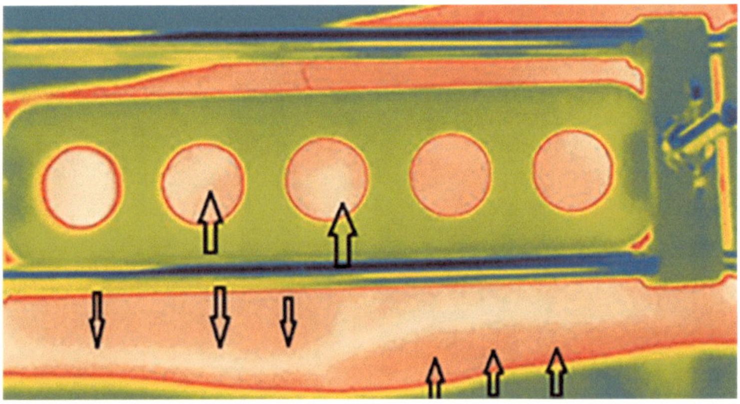

Figure 3.
Location of subcutaneous main blood vessels.

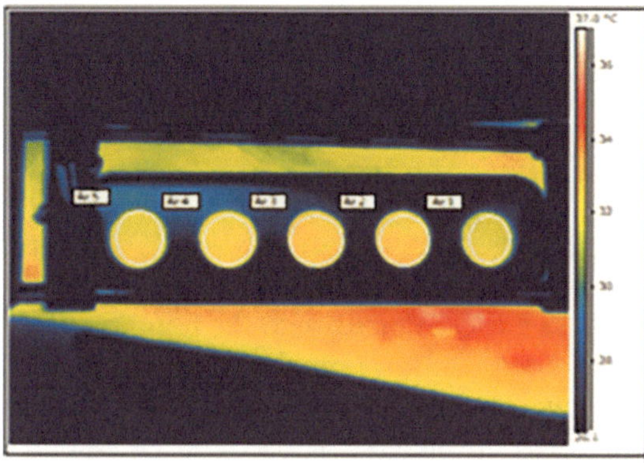

Date	21.11.2021....
Filename	Anna.jpg
Max Temperature	35.7 °C
Min Temperature	20.1 °C
Image Max. Temperature	35.7 °C
Ar1 (X_1) Average Temperature	32.0 °C
Ar2 (X_1) Average Temperature	32.9 °C
Ar3 (X_1) Average Temperature	32.9 °C
Ar4 (Neg_1) Average Temperature	32.6 °C
Ar5 (Pos_1) Average Temperature	32.4 °C

Figure 4.
First temperature imaging and determination of the temperature in the skin areas where the prick test will be performed.

X_2 – the average temperature of the skin area where the test allergen will be located after the test.

Neg_2 – the average temperature of the skin area in which the negative control will be located after the test

Pos_2 – the average temperature of the skin area in which the positive control will be located after the test (**Figure 5**).

There are different methods for analyzing the data obtained. Some authors believe that the information obtained from the second image is sufficient to determine whether a reaction is positive or negative [11]. The parameter to be analyzed is only the value of $\mathbf{X_2}$. The disadvantage of this type of analysis is that they do not include

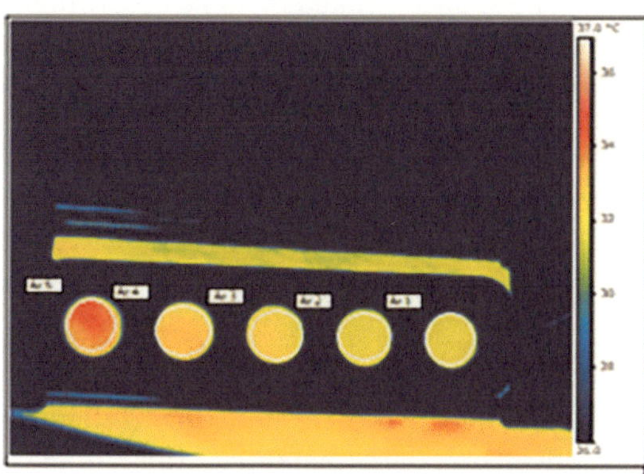

Date	21.11.2021....
Filename	Anna.jpg
Max Temperature	34.5 °C
Min Temperature	18.1 °C
Image Max. Temperature	34.5 °C
Ar1 (X_2) Average Temperature	31.9 °C
Ar2 (X_2) Average Temperature	31.9 °C
Ar3 (X_2) Average Temperature	32.5 °C
Ar4 (Neg_2) Average Temperature	32.8 °C
Ar5 (Pos_2) Average Temperature	33.8 °C

Figure 5.
Second temperature imaging and determination of the temperature in the skin areas in which the prick test was performed.

the initial skin temperature, which is individual for each patient and may affect the final assessment. The results also show large variations in different patients, so conclusions based on this indicator alone may be wrong.

Another variant found in the literature is to study the change in temperature that has occurred in the skin area of the respective allergen (Eq. (1)) [6, 11, 12].

$$\Delta X = X_2 - X_1 \tag{1}$$

When performing the test on the patient's back, an increase in the temperature of the positive reactions is reported, which reaches 2.9 degrees in the strongest reactions and is 1 degree in the other positive reactions. In our study on the volar surface of the hand, a slight rise in temperature was observed during the positive control, and in women, in some cases, even a slight cooling of the site was found. The reason for this difference is the place where the test was performed. The forearm during the test is located at a distance from the body and the temperature of her skin drops. The cooling in the negative reactions is significantly greater than the cooling that occurred in the positive control because they do not show inflammation to compensate for it. When the test is performed on the back, where the possibilities of temperature homeostasis of the body do not allow such a strong cooling within 15 minutes, a rise in temperature is observed even in negative controls [11].

The cooling of the negative reactions is more pronounced in women than in men. While in the field of positive reactions, the differences between the sexes are minimal [13].

The temperature change of the skin in the area of positive control increases in 94% of patients [14]. In such studies, this percentage reached 98% [13].

The temperature rise is significant from 1.5 to 4.00 degrees in some of the studies [6]. While in others the temperature rise in positive reactions is significantly lower: 0.9 ± 0.48 degrees [13]. The difference in results is due to the type of allergens used (local anesthetic—Mepivastesin and pollen allergens) and the intensity of the positive reactions.

Allergic inflammation in the negative reactions is absent, so the researchers did not report a change in temperature in them [8, 11, 12].

The change in temperature in most cases is indicative of the presence and strength of a positive reaction to the tested allergen. The analysis should be done according to the patient's gender, that is, variations between men and women are found.

To determine the change in temperature, which is due solely to allergic inflammation, it is necessary to take into account the drop in skin temperature during the test. It varies from patient to patient and depends on individual characteristics, such as subcutaneous tissue, blood supply, and thermal homeostasis. By changing the temperature in the area of the negative control, the cooling of the skin during the test can be monitored Eq. (2):

$$\Delta Neg = Neg_2 - Neg_1 \tag{2}$$

The difference between the value of ΔX and ΔNeg is the warming due to allergic inflammation, excluding all other factors influencing the skin temperature. Eq. (3)

$$\Delta Xa = \Delta X - \Delta Neg \tag{3}$$

The indicator ΔX has values below 0.5C for negative allergic reactions and over 0.5 for positive ones. The stronger the allergic inflammation, the higher its value. When the value is below 0.5 allergic inflammation has no clinical value. It can be considered null and void and explained by the difference in the location of the skin areas in which the test was performed. This indicator does not depend on the sex of the patient and allows a comparison of the intensity of allergic reactions between men and women [13].

Of interest are patients with sensitive skin who have dermographism. This is a condition in which the skin reacts by inflammation to nonspecific irritants. Mechanical trauma during the test elicits a response that may be incorrectly reported as allergic [15]. This condition is the reason to include positive and negative controls in the Prick test. In patients with dermographism, measurements of papules and erythema are reported and compared with those of the negative control. **Figure 6** shows three reactions—to anesthetic, to negative control, and to positive control (closest to the patient's tattoo).

All three reactions have papule sizes over 3 mm, which makes it difficult to read the standard. Thermovisiographic analysis also allows to check the temperature changes in the skin areas during the test. Four skin areas are examined (**Figures** 7 and **8**):

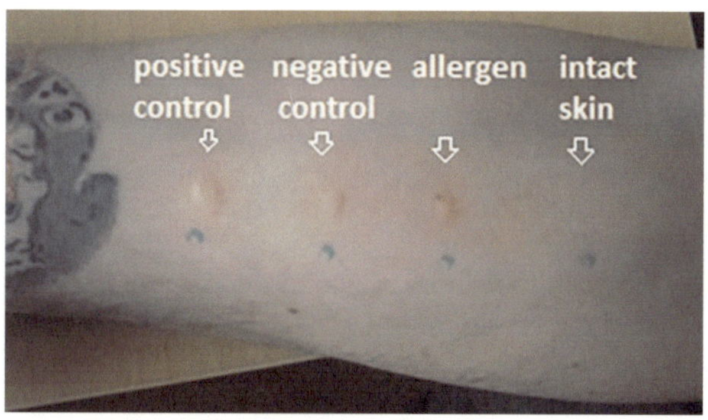

Figure 6.
Reactions after prick test of a patient with dermographism.

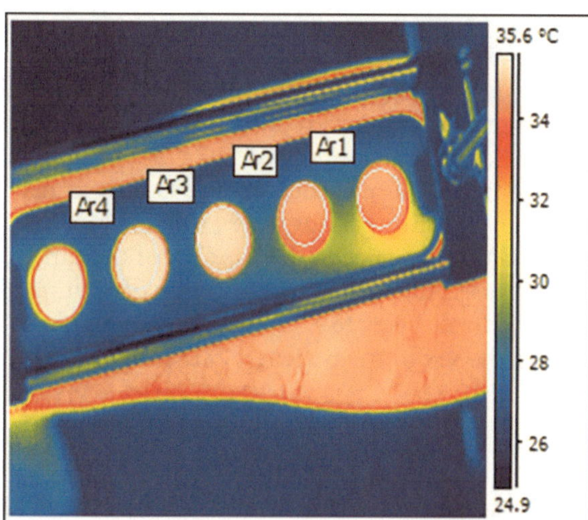

Date	20.3.2018.....
Filename	pre prick.jpg
Max Temperature	35.9 °C
Min Temperature	23.0 °C
Image Max. Temperature	35.9 °C
Ar1 ( )	34.5 °C
Ar2 (X_1)	34.5 °C
Ar3 (Neg_1)	35.1 °C
Ar4 (Pos_1)	35.3 °C

Figure 7.
First shot before prick test of a patient with dermographism.

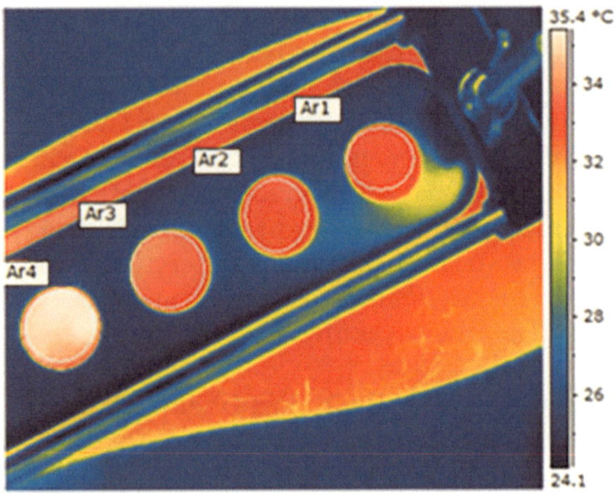

Date	20.3.2018....
Filename	post prick.jpg
Max Temperature	35.8 °C
Min Temperature	23.6 °C
Image Max. Temperature	35.8 °C
Ar1 ( )	33.5 °C
Ar2 (X_2)	33.7 °C
Ar3 (Neg_2)	34.2 °C
Ar4 (Pos_2)	35.8 °C

Figure 8.
Second imaging after prick test of a patient with dermographism.

Ar1 – intact skin area in the area of which there is no legal test.

Ar2 – examination of a local anesthetic (mepivacaine).

Ar3 – negative control.

Ar4 – positive control.

After analyzing the main indicators, the additional ones can be calculated by Eqs. (1)–(3):

$\Delta X = X_2 - X_1$ = 33.7–34.5 = -0.8.

$\Delta Neg = Neg_2 - Neg_1 = 34.2\text{–}35.1$ = -0.9.

$\Delta Xa = \Delta X - \Delta Neg$ = -0.8 – (-0.9) = 0.1.

The results show that there is no evidence of allergy to the studied local anesthetic. If we make similar calculations for the positive control, we will get a value of $\Delta Xa = 1.4$.

$\Delta Xa = (Pos_2 - Pos_1) - \Delta Neg$ = (35.8–35.3) – (-0.9) = 1.4.

The calculations show that the reactions of dermographism are subject to thermal imaging analysis. This is a major advantage over standard reporting in the presence of dermographism. More studies of this type of patient are needed.

The analysis of the results obtained from the thermal imaging provides information about the temperature changes in the skin at the places where the test was performed. By considering various parameters, a better understanding of the inflammation at the sites of allergens and controls is achieved. The calculation of additional parameters based on the results of both controls helps to unify the process and to create limit values that do not depend on the sex of the patient and the individual characteristics of his/her skin.

3. Thermal imaging analysis of skin-allergic reactions in intradermal tests

The test has many similarities with the Prick test—it is performed on the patient's forearm, areas with superficial blood vessels are avoided. The hand is placed on a stable base. The allergen is administered intradermally using a syringe and needle. The amount is about 0.02 ml. A papule of about 2–3 mm forms on the skin above the tip of the needle. In addition to the tested allergens, the test is performed with negative and positive control. In this test, it is necessary to avoid areas with superficial main blood vessels. Unlike the Prick test, here the number of allergens is usually significantly higher. Thermal imaging is done after the test. Due to the large number of tests located close to each other, the use of the Prick test stand is not appropriate. On this trichina, fine metal indicators are placed on the hand, which are clearly visible on the thermal image and indicate the place where the test was performed (**Figure 9**).

The thermal imaging image is similar to that of the Prick test, but due to the lack of an armrest with outlined areas, their size and location are determined subjectively depending on the course of the isotherms (**Figure 10**).

Analysis of the temperature of the skin areas AFTER performing an intradermal test.

The survey provides information on the main temperature parameters:

Z – the average temperature of the skin area in which the test allergen is located.

Neg – the average temperature of the skin area in which the negative control is located

Pos – the average temperature of the skin area in which the positive control is located (**Figure 11**)

As in the Prick test, the analysis here can be based on the absolute temperature value of the reactions [16]. Another approach is to find the difference between the

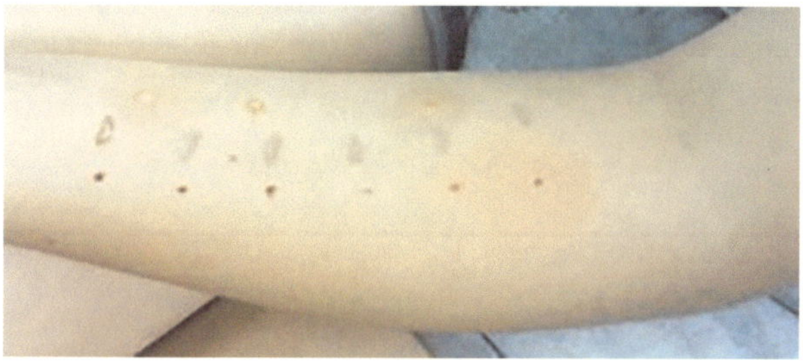

Figure 9.
Intradermal allergy test.

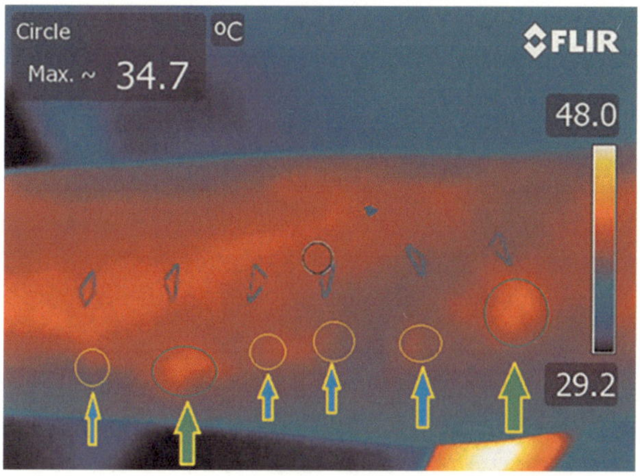

Figure 10.
Thermal imaging image after the intradermal test. Green arrows indicate positive reactions and blue arrows indicate negative reactions.

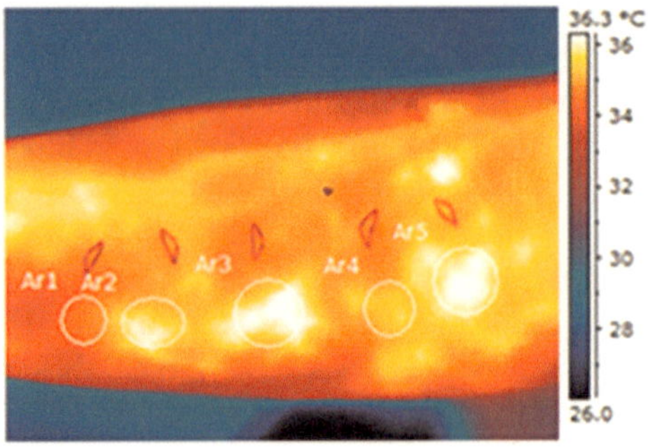

Date	18.6.2019.....
Filename	IR_6873.jpg
Max Temperature	36.6 °C
Min Temperature	25.8 °C
Image Max. Temperature	36.6 °C
Ar1 Average Temperature	34.5 °C
Ar2 Average Temperature	35.6 °C
Ar3 Average Temperature	35.9 °C
Ar4 Average Temperature	35.4 °C
Ar5 Average Temperature	36.0 °C

Figure 11.
Outline the zones of reactions on the thermal image.

reaction temperature and that of the negative control. The value of allergic inflammation is measured as in the Prick test, but the control area is replaced with the area of negative control (Eq. (4)).

$$\Delta Z = Z - Neg \tag{4}$$

At values of ΔZ above 0.6 degrees, the reactions can be considered positive and below 0.6 degrees - negative [13]. A similar comparison can be made with respect to the positive control (Eq. (5))

$$\Delta Z = Z - Pos \tag{5}$$

With the values of ΔZ above -1.0 degrees, the reactions can be considered positive and below -1 degree - negative [13].

In intradermal tests, needle pricking results in severe mechanical trauma that causes nonallergic inflammation at the puncture site. For this reason, comparing the changes in temperature in each of the skin areas is not as informative as in the Prick test. Also, the comparison of the temperature of the reactions with those of the negative and positive control gives sufficient information not only about whether the reaction is positive but also what is its intensity. With the help of both controls, the reactivity of the skin can be determined in the absence of allergic inflammation and in case of a strong skin-allergic reaction. The stronger positive reactions show closer temperature values to those of the positive control, while the weaker ones approach more to the negative control.

4. Thermal imaging analysis of skin-allergic reactions after patch test

In this type of test, the allergens are placed on a hypoallergenic sticker with chambers, which is glued to the patient's back and remains there for 48 hours. The long period of time presupposes a different approach in the thermal imaging analysis—a photograph is taken of the patient's skin after the patch has been removed and the irritating reaction has passed since its removal. The problem is that the test is done on a large area of skin with underlying muscle groups, tendons, and vertebrae. Like the volar surface of the forearm, the temperature is different in different areas. This makes the use of negative control (empty chamber without hapten in it) unsuitable for temperature analysis, due to the large distance at which it is located relative to allergens. For example, if the empty chamber is located near the spine, the skin temperature in that area will always be lower than the temperature in an area above a muscle, regardless of the presence of an allergic reaction. On the other hand, unlike the Prick test, here the reactions are limited only to the size of the area with which the allergen has been in contact and do not affect **the surrounding skin areas** [17, 18]. Therefore, the correct approach is to compare the temperature of each reaction with the temperature of a nearby skin area that has not been in contact with the allergen.

A template is placed to identify the skin areas that have been in contact with the bitten. With its help, the location of each reaction can be clearly determined, both in standard and thermal imaging (**Figure 12**).

The size of the templates may vary depending on the number of haptens tested (**Figure 13**).

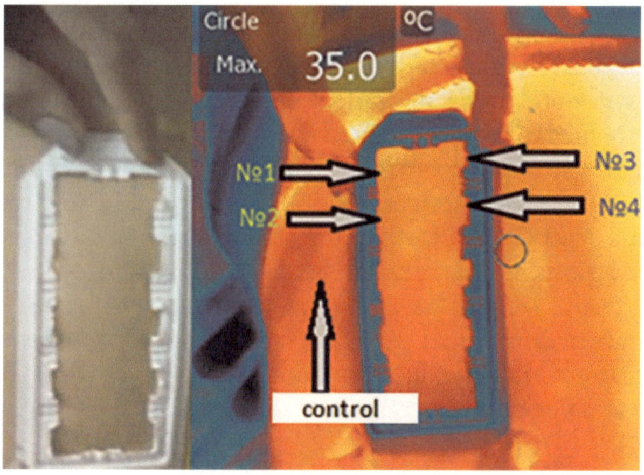

Figure 12.
Thermal imaging of a patient's back after a patch test with a template for thermovisiographic reading. The skin areas of the four allergens and the lateral area used for the control area are marked with arrows.

Figure 13.
Investigated reactions and their respective control areas.

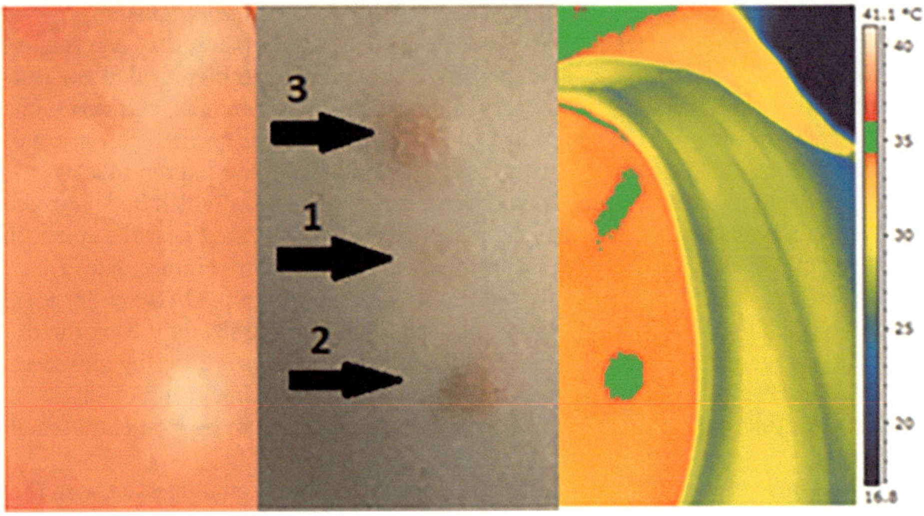

Figure 14.
Different positive reactions (standard and thermal imaging). 1 - slightly (weak) positive; 2 - strongly positive; 3 - extremely positive.

The standard reading divides the established reactions into several groups—negative, doubtful, weakly positive, strongly positive, and extremely positive (**Figure 14**).

In both the Prick test and the Patch test, different approaches can be used to analyze reactions. Some studies have involved measuring the absolute skin temperature of reactions without comparing control sites. When comparing the average values of the reactions, the negative ones have an average temperature of 34.7°C; weakly positive - 35.1°C, and strongly positive - 35.7°C. At the maximum temperatures, the values are—for the negative reactions - 35.0°C; in the weakly positive - 35.5°C, in the **strongly positive - 36.0°C** [18]. Despite the seemingly large differences in the values of the different reactions in practice, it is established that the same allergic inflammations can be manifested with different skin temperatures [19]. The main reason for this is the temperature differences on the surface of the skin on the patient's back. Often there are many nonallergic inflammations on the skin, which can be mistaken for positive and even very positive reactions (**Figure 15**).

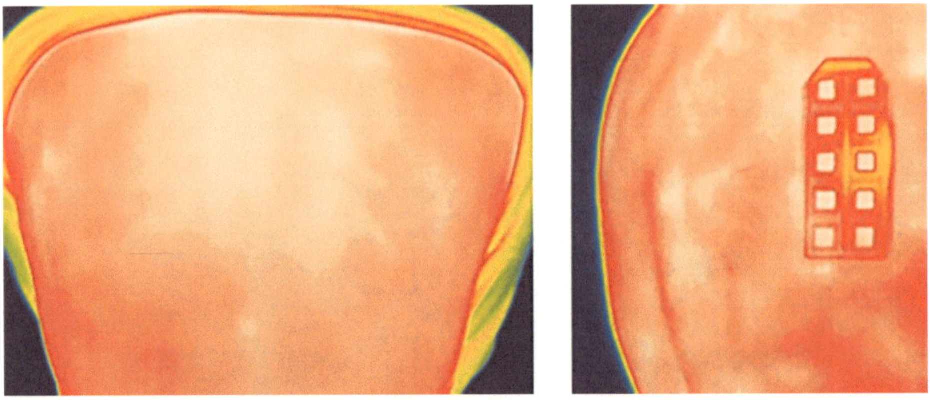

Figure 15.
Thermal imaging of the back without skin inflammations and thermovision imaging of the back with nonallergic skin inflammations on which a patch test was performed.

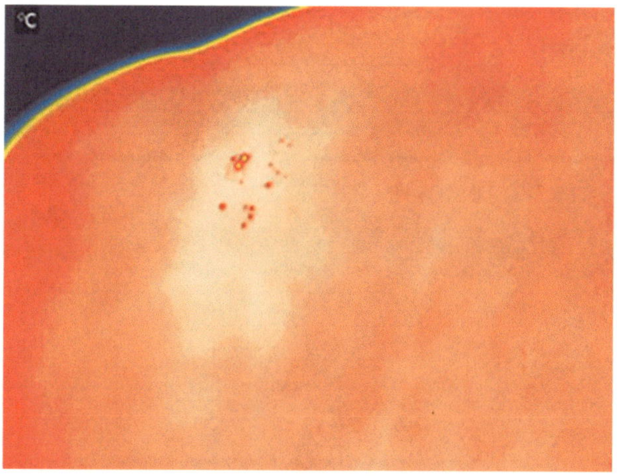

Figure 16.
Extremely strong positive reaction with vesicles in three of the studied haptens.

Another approach is to analyze the difference between the skin temperature at the reaction and that of the corresponding control area. To determine the temperature, which is due solely to allergic inflammation ΔY, we subtract the temperature of the control area $Y_{control}$ from the temperature in the skin area that was in contact with the hapten $Y_{reaction}$ (Eq. (6)).

$$\Delta Y = Y_{reaction} - Y_{control} \tag{6}$$

The indicator ΔY with values above 1.6 is considered extremely positive. Values between 0.9 and 1.6 are characteristic of strongly positive reactions. It is difficult to distinguish the weakly positive reactions from the suspicious and negative reactions because in all three ΔY, it is below 0.9 degrees. The role of calorimetry in the Patch test is to strongly distinguish between highly positive and extremely positive reactions and to confirm the results of the standard reading.

Of particular interest are the extremely positive reactions in which vesicles form in the center of the skin reaction. They can be easily recognized by the thermal image—there is an area with a temperature higher than the surrounding above 1.6 degrees and in its center a few points with a significantly lower temperature corresponding to the vesicles. The bubbles that form are filled with serous fluid, which cools much more easily than the skin and therefore its temperature is much lower (**Figure 16**).

There is no difference in the thermal imaging test in the Patch test for early and late reading. The analysis may include both mean and maximal skin temperature. Of great importance is the choice of control site for each of the reactions. This should be an area close to the reaction area, but not exposed to a hapten. In patients with irritated skin on the back, an area where there is less inflammation should be chosen. This supports both standard and thermal imaging of the results.

5. Conclusion

The measurement of the temperature of the skin areas in allergic tests gives objective data on the inflammation and its intensity. Each test examines different

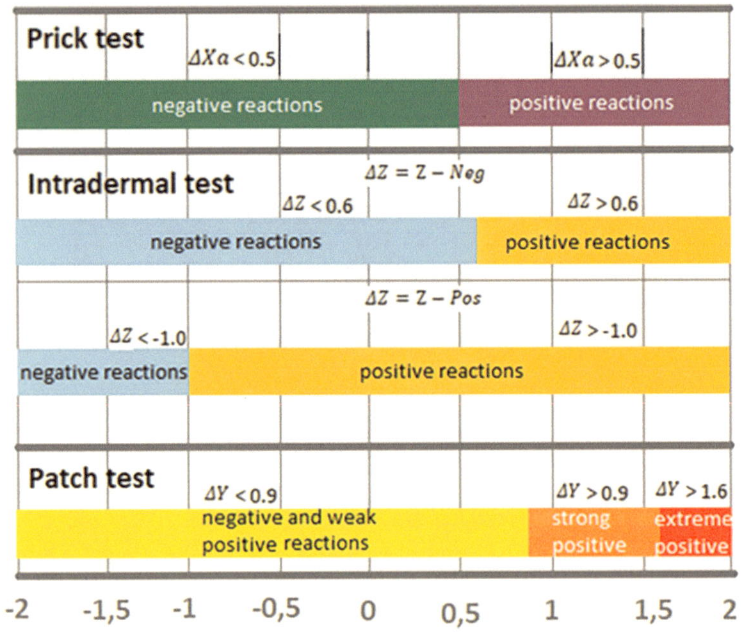

Figure 17.
Temperature indicators used in the various allergy tests and their limit values.

indicators, the value of which complements the results of standard reporting and assists clinicians in their practice (**Figure 17**).

In the Prick test, reactions with a value of показа ΔXa below 0.5 are negative, while in higher values, they should be considered positive.

In intradermal tests, reactions with показа ΔZ (negative control) below 0.6 were negative, while at higher values, they were positive.

In intradermal tests, reactions with показа ΔZ (positive control) below -1.0 were negative, while at higher values, they were positive.

In Patch test, reactions with показа ΔY below 0.9 are negative or weakly positive; over 0.9 - strongly positive; over 1.6 - extremely positive.

The studies presented in this article have a high clinical value and provide a more complete understanding of the results of allergy tests. The high correlation between the clinical results and the thermal imaging examination (over 95% correlation) are the grounds for applying the method in parallel with the other diagnostic methods in allergology.

Conflict of interest

The authors declare no conflict of interest.

Author details

Evgeni Stanev* and Maria Dencheva
Faculty of Dental Medicine, Sofia, Bulgaria

*Address all correspondence to: stanev242@gmail.com

References

[1] Petrunov B, Dimitrov V, Kiselova-Yaneva A. Clinical immunology. Clinical allergology. Dental clinical allergology (Клинична имунология. Клинична алергология. Дентална клинична алергология.). 1st ed. Bulgaria: Apco; 2012

[2] Chemotechnique MB Diagnostics AB. Interpretation of patch test results [Internet]. Available from: https://www.chemotechnique.se/patch-testing/interpretation-/ [Accessed: 2021-12-18]

[3] Johansen JD, Aalto-Korte K, Agner T, Andersen KE, Bircher A, Bruze M, et al. European Society of Contact Dermatitis guideline for diagnostic patch testing–recommendations on best practice. Contact Dermatitis. 2015;**73**(4):195-221. DOI: 10.1111/cod.12432

[4] Frati F, Incorvaia C, Cavaliere C, Di Cara G, Marcucci F, Esposito S, et al. The skin prick test. Journal of Biological Regulators and Homeostatic Agents. 2018;**32**(1 Suppl. 1):19-24

[5] Fernández-Cuevas I, Marins JCB, Lastras JA, Carmona PMG, Cano SP, García-Concepción MÁ, et al. Classification of factors influencing the use of infrared thermography in humans: A review. Infrared Physics & Technology. 2015;**71**:28-55. DOI: 10.1016/j.infrared.2015.02.007

[6] Rokita E, Rok T, Tatoń G. Evaluation of allergic response using dynamic thermography. Opto-Electronics Review. 2015;**23**(1):55-61. DOI: 10.1515/oere–2015–0007

[7] De Weck AL, Derer T. Critical evaluation of the use of skin tests and cellular tests in standardization of allergens. Arb Paul Ehrlich Inst Bundesamt Sera Impfstoffe Frankf A M. 1994;**87**:89-114

[8] Rok T, Rokita E, Tatoń G, Guzik T, Śliwa T. Thermographic assessment of skin prick tests in comparison with the routine evaluation methods. Postępy Dermatologii i Alergologii. 2016;**33**(3): 193. DOI: 10.5114/ada.2016.60611

[9] Stanev E, Dencheva M, Lyapina M, Forghani P. Thermographic examination of prick test reactions with local anesthetic. Journal of Thermal Analysis and Calorimetry. 2020;**140**(1): 225-231. DOI: 10.1007/s10973-019-08814-4

[10] Heinzerling L, Mari A, Bergmann KC, Bresciani M, Burbach G, Darsow U, et al. The skin prick test – European standards. Clinical and Translational Allergy. 2013;**3**(1):3. DOI: 10.1186/2045-7022-3-3

[11] Phipatanakul CS, Slavin RG. Use of thermography in clinical allergy. The Journal of Allergy and Clinical Immunology. 1972;**50**:264-275. DOI: 10.1016/0091-6749(72)90025-5

[12] Huang HW, Horng TL. Bioheat transfer and thermal heating for tumor treatment. In: Heat Transfer and Fluid Flow in Biological Processes. Academic Press; 2015. pp. 1-42. DOI: 10.1016/B978-0-12-408077-5.00001-8. Available from: https://tlhorng.math.fcu.edu.tw/paper/chap1.pdf, https://www.sciencedirect.com/science/article/pii/B9780124080775000018

[13] Stanev E. Thermovision in skin-allergic tests [thesis]. Medical University of Sofia; 2020

[14] Bagnato GF, Gulli S, Denuzzo G, Di Cesare E, Sansotta C, D'Ambrosio FP, et al. Measurement of allergen-induced skin reactions by computerized dynamic telethermography. Journal of

Investigational Allergology & Clinical Immunology. 1997;7:238-241

[15] Schoepke N, Młynek A, Weller K, Church MK, Maurer M. Symptomatic dermographism: An inadequately described disease. Journal of the European Academy of Dermatology and Venereology. 2015;**29**(4):708-712. DOI: 10.1111/jdv.12661

[16] Koyama N, Hirata K, Hori K, Dan K, Yokota T. Computer-assisted infrared thermographic study of axon reflex induced by intradermal melittin. Pain. 2000;**84**(2–3):133-139. DOI: 10.1016/S0304-3959(99)00192-X

[17] Anzengruber F, Alotaibi F, Kaufmann LS, Ghosh A, Oswald MR, Maul JT, et al. Thermography: High sensitivity and specificity diagnosing contact dermatitis in patch testing. Allergology International. 2019;**68**(2): 254-258. DOI: 10.1016/j.alit.2018.12.001

[18] Станев Е, Денчева М. Термовизиографско изследване на кожни реакции след Patch тест. ПРОБЛЕМИ НА ДЕНТАЛНАТА МЕДИЦИНА PROBLEMS OF DENTAL MEDICINE, 62

[19] Dencheva M, Lyapina M, Kisselova-Yaneva A, Garov A, Hristova S, Konstantinova M, et al. Thermovision in dental allergology. J IMAB – Annual Proceeding Scientific Papers. 2014;**20**(3):558-562. DOI: 10.5272/jimab.2014203.558